AI AWARENESS SERIES

AI in Project Management

Daniel Rockbridge

Contents

Introduction

The integration of artificial intelligence into project management represents one of the most significant transformations in how we plan, execute, and deliver projects in the modern era. As organizations face increasingly complex challenges, tighter deadlines, and the need for data-driven decision-making, AI emerges not as a replacement for human project managers but as a powerful ally that amplifies their capabilities.

This book serves as a comprehensive guide for project management professionals navigating the AI revolution. Whether you're a seasoned project manager looking to enhance your toolkit, a PMO leader seeking to transform organizational practices, or a forward-thinking professional preparing for the future of work, you'll find practical insights and actionable strategies within these pages.

The journey begins with understanding the fundamental AI technologies reshaping our field—from machine learning and natural language processing to predictive analytics and generative AI. We then explore how these technologies are redefining the project manager's role, transforming it from administrative coordination to strategic leadership augmented by intelligent systems.

Through detailed examinations of AI applications across the entire project lifecycle—from initiation through closure—you'll discover how artificial intelligence enhances every aspect of project management. We delve into real-world tools, examine predictive analytics capabilities,

Introduction

and explore conversational AI assistants that are already improving project outcomes across industries.

But technology alone doesn't drive transformation. This book addresses the critical human elements: building an AI-ready culture, managing risks and governance, and preparing your organization for sustainable change. We tackle the challenges head-on, from data quality concerns to ethical considerations, providing frameworks for responsible AI adoption.

Looking ahead, we examine emerging trends and evolving standards that will shape the future of AI-enhanced project management, including the potential for autonomous project systems and the integration of sustainability metrics through intelligent automation.

Each chapter combines theoretical foundations with practical applications, offering not just concepts but concrete strategies you can implement immediately. The goal is not to predict a distant future but to equip you with the knowledge and tools needed to thrive in the AI-augmented present while preparing for continued evolution.

As we stand at this intersection of human expertise and artificial intelligence, the opportunity before us is unprecedented. AI doesn't diminish the importance of project managers—it elevates their potential to deliver extraordinary results through the powerful combination of human judgment and machine intelligence.

Chapter 1: Understanding AI Technologies

Artificial Intelligence represents the simulation of human intelligence processes through machines, enabling them to perform tasks autonomously without constant human intervention. The learning component is fundamental - AI systems continuously improve their performance by analyzing vast amounts of data over time, identifying patterns and relationships that might escape human observation. The reasoning and self-correction aspects distinguish AI from simple automation. These systems can make decisions based on available information, evaluate the outcomes of those decisions, and adjust their approach based on feedback and new data. This creates a dynamic learning environment where AI systems become more effective and accurate through experience, making them invaluable tools for complex problem-solving across industries.

AI systems are categorized by capability into three main levels. Narrow AI, which we use today, excels at specific tasks like image recognition or language translation but cannot transfer knowledge between domains. General AI, still theoretical, would match human cognitive abilities across all areas. Superintelligent AI represents hypothetical systems surpassing human intelligence entirely. Beyond capability, we classify AI by approach: reactive machines respond to current situations without memory, limited memory systems learn from recent data, theory of mind AI would understand emotions and intentions, and self-aware AI would possess consciousness. Understanding these classifications helps businesses choose appropriate AI solutions and set realistic expectations for implementation timelines and capabilities.

Chapter 1: Understanding AI Technologies

The evolution of AI includes several pivotal milestones that shaped today's landscape. The Turing Test, proposed by Alan Turing, established the benchmark for machine intelligence by testing whether humans can distinguish between human and machine responses in conversation. Expert systems marked early practical AI applications, using rule-based decision-making to solve specific domain problems in fields like medical diagnosis and financial analysis. Deep learning emergence revolutionized AI by introducing multi-layered neural networks capable of processing complex patterns in data, enabling breakthroughs in image recognition and natural language processing. Recent advances in neural networks and reinforcement learning allow machines to learn from experience, continuously improving performance through trial and error, much like human learning processes.

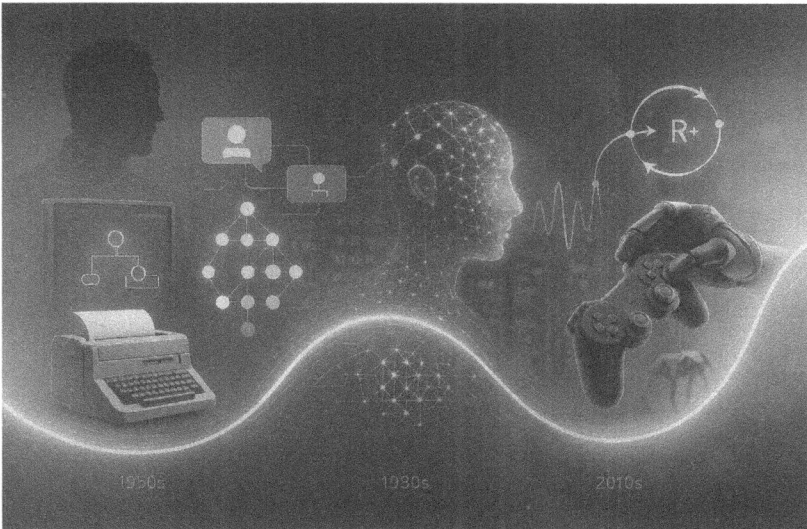

Chapter 1: Understanding AI Technologies

Machine learning represents the core methodology enabling AI systems to improve automatically through experience. The learning process begins with data - vast datasets containing examples, patterns, and relationships that algorithms can analyze and understand. During the training phase, models process labeled data to identify underlying patterns and relationships, creating mathematical representations that can predict outcomes for new, unseen data. The validation and testing phases ensure models generalize effectively beyond their training data, preventing overfitting and ensuring reliability in real-world applications. This systematic approach transforms raw data into actionable insights, enabling businesses to make data-driven decisions with confidence. The iterative nature of machine learning means models continuously improve as more data becomes available.

Machine learning encompasses three primary learning paradigms, each suited for different problem types. Supervised learning uses labeled training data to predict outcomes, making it ideal for classification tasks like email spam detection or regression problems like sales forecasting. The algorithm learns from input-output pairs to make accurate predictions on new data. Unsupervised learning discovers hidden patterns in unlabeled data, identifying customer segments, anomalies, or underlying data structures without predefined categories. This approach reveals insights that might not be obvious through traditional analysis. Reinforcement learning trains agents through rewards and penalties, optimizing decision-making over time. This approach excels in dynamic environments where the algorithm must balance immediate rewards with long-term objectives, making it perfect for applications like game playing or resource optimization.

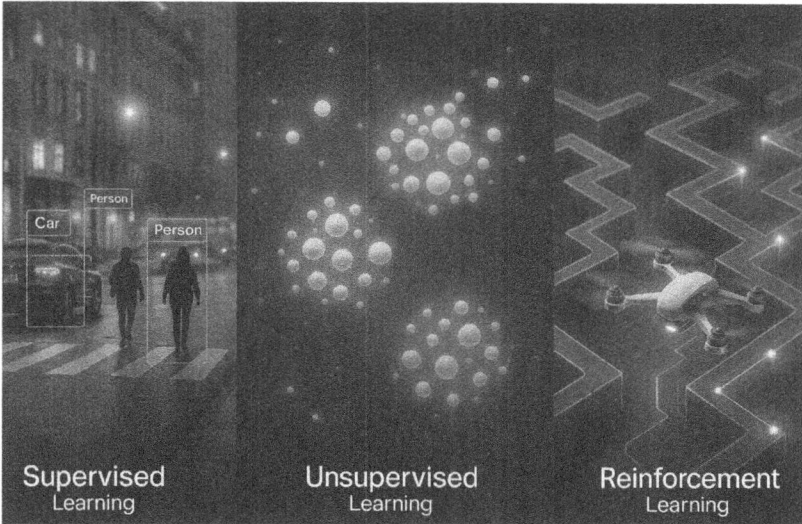

Machine learning applications span numerous industries, demonstrating its versatility and practical value. Image recognition systems accurately identify and classify visual content, enabling applications from medical diagnosis to autonomous vehicles. These systems can process thousands of images in seconds, identifying patterns and anomalies that might escape human observation. Fraud detection algorithms analyze transaction patterns to identify suspicious activities in real-time, protecting both businesses and consumers from financial crimes. Recommendation engines power personalized experiences across platforms, analyzing user behavior and preferences to suggest relevant products, content, or services. Speech recognition technologies convert spoken language into text, enabling voice assistants and automated transcription services. Autonomous vehicles rely on machine learning for navigation, obstacle detection, and decision-making in complex traffic environments.

Chapter 1: Understanding AI Technologies

Natural Language Processing focuses on three fundamental linguistic concepts that enable machines to understand and generate human language. Syntax analysis examines grammatical structure, identifying how words relate to each other within sentences to understand meaning and context. This involves parsing sentence structure, identifying parts of speech, and understanding grammatical relationships that provide the foundation for language comprehension. Semantics understanding goes deeper, focusing on the actual meaning of words and sentences, interpreting context and intent beyond literal definitions. This includes understanding metaphors, implied meanings, and contextual nuances. Pragmatics interpretation studies how language functions in real-world contexts, considering speaker intention, cultural factors, and situational elements that influence meaning. Together, these concepts enable AI systems to process human language with increasing sophistication and accuracy.

NLP employs several sophisticated techniques to analyze text and speech effectively. Tokenization breaks text into manageable units like words or phrases, while parsing analyzes grammatical structure to understand relationships between these units. This preprocessing step is crucial for all subsequent analysis. Sentiment analysis determines emotional tone and opinion within text, enabling businesses to understand customer feedback, social media sentiment, and market perception. This capability is invaluable for brand monitoring and customer service optimization. Speech recognition converts spoken language into text, enabling voice-controlled interfaces and automated transcription services. Named Entity Recognition extracts key information like names, locations, dates, and organizations from text, enabling automated information extraction and data organization. These techniques work together to transform unstructured text into structured, actionable information.

NLP applications demonstrate practical value across multiple business functions. Intelligent chatbots leverage NLP to provide sophisticated customer service, understanding customer queries and providing relevant responses in real-time. These systems can handle routine inquiries, escalate complex issues, and maintain conversation context throughout interactions. Language translation tools powered by NLP break down communication barriers globally, enabling real-time translation that preserves context and nuance. Modern translation systems understand idioms, cultural references, and domain-specific terminology. Sentiment analysis applications monitor public opinion, social media trends, and customer feedback, providing businesses with insights into brand perception and market sentiment. This enables proactive customer service, product development guidance, and reputation management strategies that respond to actual customer needs and concerns.

Generative AI represents a paradigm shift from analytical to creative AI capabilities. These systems create entirely new content by learning patterns from existing data, going beyond simple analysis to generate original outputs. The learning process involves analyzing massive datasets to understand underlying patterns, relationships, and structures within the training data. This deep understanding enables the system to generate new content that maintains consistency with learned patterns while creating original variations. Generative AI can produce diverse content types including text, images, music, code, and multimedia content across numerous creative domains. The sophistication of these systems continues to advance, enabling increasingly realistic and contextually appropriate content generation. This capability opens new possibilities for creative industries, content marketing, product development, and automated content creation at scale.

Large Language Models represent the cutting edge of generative AI, trained on vast datasets encompassing billions of text examples from books, articles, websites, and other written sources. This extensive training enables LLMs to learn diverse language patterns, cultural contexts, writing styles, and domain-specific knowledge across virtually every field of human knowledge. Deep neural networks process this information through multiple layers, each identifying increasingly complex patterns and relationships within the data. The result is human-like language generation that can adapt to context, maintain conversation flow, understand nuanced instructions, and generate coherent responses across countless topics. LLMs excel at conversational interactions, creative writing, technical documentation, and problem-solving applications, making them valuable tools for education, customer service, content creation, and professional assistance across industries.

Generative AI applications are transforming multiple business functions through automation and enhancement of creative processes. Automated marketing content creation enables businesses to generate blog posts, social media content, email campaigns, and advertising copy at scale while maintaining brand consistency and personalization. Code writing assistance helps developers by generating snippets, suggesting optimizations, debugging code, and even creating entire functions based on natural language descriptions, significantly accelerating development cycles. AI-powered chatbots provide sophisticated customer interaction capabilities, handling complex queries, maintaining context across conversations, and offering personalized recommendations based on customer history and preferences. Personalized recommendation systems analyze user behavior, preferences, and historical data to suggest relevant products, content, or services, improving user experience and driving engagement across platforms and applications.

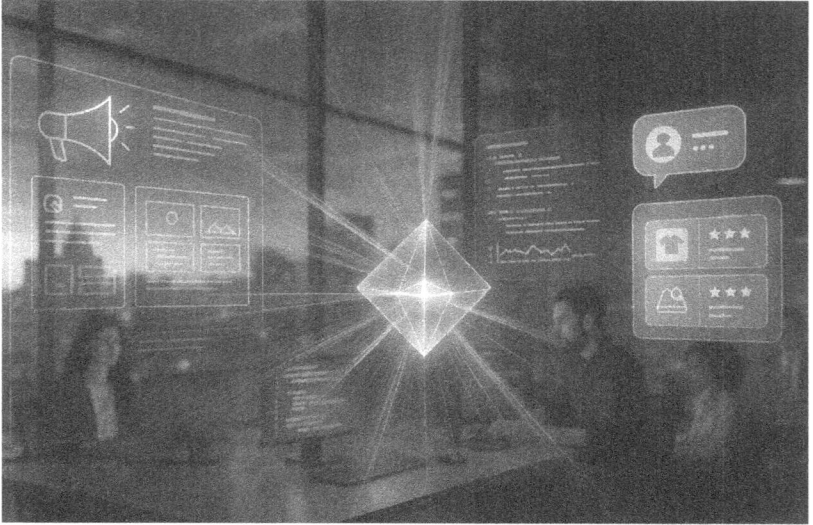

Predictive analytics forms the foundation of data-driven decision making by leveraging historical data to forecast future events and trends. The process begins with comprehensive historical data analysis, identifying patterns, seasonal trends, correlations, and anomalies that might influence future outcomes. Machine learning algorithms process this historical information to build sophisticated models capable of forecasting upcoming events with measurable accuracy levels. These models continuously improve as new data becomes available, refining predictions and adapting to changing conditions. Applications include demand forecasting for inventory management, risk assessment for financial services, customer behavior prediction for marketing optimization, and maintenance scheduling for operational efficiency. Predictive analytics enables businesses to anticipate challenges, optimize resource allocation, and make informed strategic decisions based on data-driven insights rather than intuition alone.

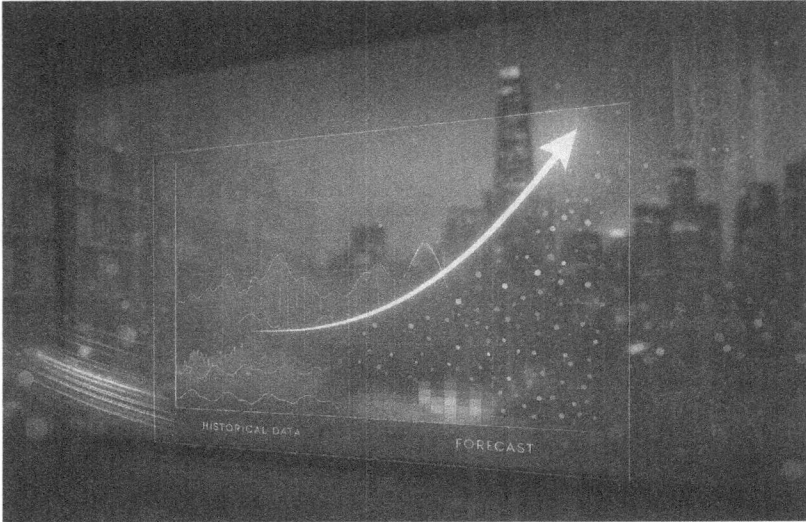

Prescriptive AI advances beyond prediction to recommend specific actions for achieving desired outcomes. Rather than simply forecasting what might happen, prescriptive AI suggests what should be done to optimize results. This approach relies heavily on optimization algorithms that evaluate multiple scenarios, constraints, and objectives to identify the best possible decisions under given circumstances. Simulations play a crucial role, enabling prescriptive AI to model various outcomes and refine strategies for achieving desired results. The system considers multiple variables, potential obstacles, resource limitations, and strategic objectives to provide actionable recommendations. This capability is particularly valuable in complex decision-making scenarios where multiple factors must be balanced, such as supply chain optimization, financial portfolio management, resource allocation, and strategic planning across various business functions.

The relationship between predictive and prescriptive AI creates a powerful decision-making framework for businesses. Predictive AI analyzes historical and current data to forecast possible future events, providing the foundation for understanding what might happen under various scenarios. This forecasting capability informs risk assessment, trend analysis, and probability calculations essential for strategic planning. Prescriptive AI builds upon these predictions to recommend optimal actions based on predicted outcomes, business objectives, and available resources. The combined impact creates a comprehensive decision-making system that not only anticipates future challenges and opportunities but also provides specific guidance for navigating complex business environments. Together, these approaches enhance decision-making quality in dynamic, uncertain environments where traditional analysis methods might fall short of providing actionable insights.

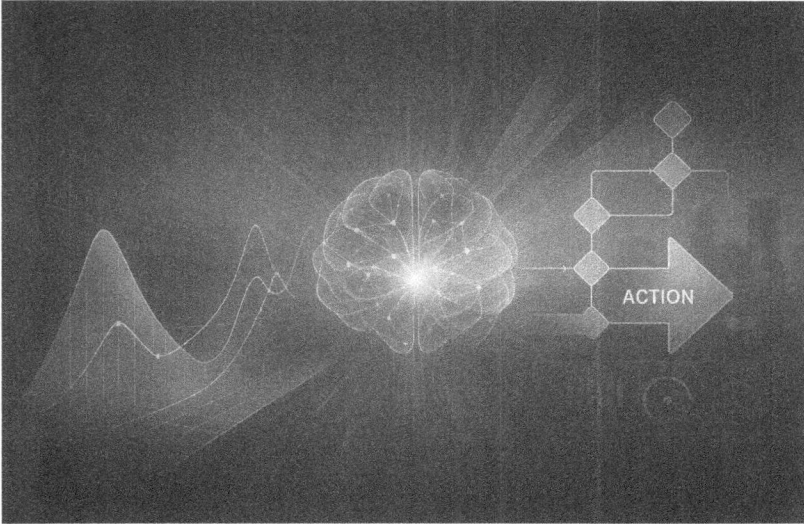

AI integration in SaaS platforms brings significant benefits while introducing new challenges that organizations must carefully manage. Automation enhancement streamlines workflows by eliminating repetitive manual tasks, enabling employees to focus on higher-value strategic activities while improving operational efficiency and reducing human error. Personalization and analytics capabilities provide users with customized experiences and advanced data insights, enabling better decision-making through real-time performance monitoring and predictive analytics. However, data privacy concerns arise as AI systems require access to sensitive information, necessitating robust security measures, compliance protocols, and transparent data handling practices. Complexity and cost considerations include implementation challenges, training requirements, system integration needs, and ongoing maintenance expenses that must be weighed against potential benefits and ROI expectations.

Chapter 1: Understanding AI Technologies

AI-powered SaaS platforms demonstrate practical applications across core business functions. Predictive sales analytics within CRM systems analyze historical sales data, customer interactions, and market trends to forecast sales performance, identify high-probability prospects, and optimize resource allocation for maximum revenue generation. These systems help sales teams prioritize leads, predict customer lifetime value, and identify upselling opportunities. Automated customer support platforms utilize AI-driven chatbots and virtual assistants to provide instant responses to common queries, escalate complex issues appropriately, and maintain conversation context throughout customer interactions. Marketing automation tools leverage AI to optimize campaign performance, personalize customer engagement across multiple channels, segment audiences based on behavior and preferences, and automatically adjust messaging based on real-time performance data and customer responses.

Future directions for AI in software services promise significant enhancements across user experience and operational capabilities. Enhanced user experiences will emerge through personalized interfaces that adapt to individual preferences, automated task completion, and intuitive interactions that reduce learning curves and improve productivity. Real-time insights will become standard, providing immediate data analytics and actionable intelligence to support quick decision-making in rapidly changing business environments. Adaptive systems will dynamically adjust software behavior based on evolving user needs, environmental changes, and performance feedback, creating more responsive and intelligent applications. Business function integration will seamlessly connect AI capabilities across departments, enhancing collaboration, data sharing, and operational efficiency while breaking down traditional silos that limit organizational effectiveness and innovation potential.

In conclusion, the rapid evolution of AI technologies continues transforming industries worldwide at an unprecedented pace. From machine learning and natural language processing to generative AI and predictive analytics, these technologies offer wide-ranging impact across business functions, enhancing innovation, operational efficiency, and competitive advantage. Understanding AI concepts and applications is crucial for leveraging these technologies' potential benefits while navigating associated challenges responsibly. Organizations that embrace AI thoughtfully, with proper planning and implementation strategies, will be better positioned to succeed in an increasingly AI-driven business landscape. The key lies in understanding both capabilities and limitations, ensuring ethical implementation, and maintaining focus on human-centered value creation through intelligent technology adoption.

Chapter 2: Evolving Role of the Project Manager

AI is fundamentally reshaping traditional project management responsibilities across four critical dimensions. First, automation handles routine tasks like scheduling updates, progress tracking, and basic reporting, freeing managers for strategic work. Second, this liberation allows increased focus on strategic decisions, complex problem-solving, and innovation initiatives. Third, managers now oversee hybrid teams where AI systems work alongside human team members, requiring new collaboration frameworks. Finally, AI tools enhance stakeholder engagement through automated communication, real-time updates, and personalized reporting. These changes represent a shift from administrative task management to strategic orchestration and human-AI collaboration facilitation.

Technology advancement creates both challenges and opportunities for traditional project management approaches. Conventional methods built on manual processes, paper-based tracking, and human-only decision-making face disruption from intelligent systems. However, this disruption opens doors for enhanced efficiency and accuracy. Project managers now integrate AI tools for planning optimization, predictive scheduling, and continuous monitoring capabilities. The key lies in understanding that AI doesn't replace project management fundamentals but amplifies their effectiveness. Modern project managers must embrace this integration, learning to leverage AI capabilities while maintaining essential human judgment, creativity, and leadership skills that technology cannot replicate.

Chapter 2: Evolving Role of the Project Manager

The AI transformation presents both significant opportunities and notable challenges for project managers. AI-driven insights and automation offer unprecedented efficiency gains, deeper data analysis, and enhanced decision-making capabilities. These tools can process vast amounts of project data, identify patterns humans might miss, and automate routine administrative tasks. However, challenges include effectively integrating AI systems into existing workflows without disruption, ensuring team alignment when introducing new technologies, and addressing ethical concerns about AI decision-making. Success requires balancing AI capabilities with human oversight, maintaining team cohesion during technological transitions, and establishing clear ethical guidelines for AI use in project environments.

AI transforms project insights through sophisticated data analysis capabilities that far exceed human capacity for processing complex information. AI systems can analyze massive datasets from multiple project sources, identifying meaningful patterns, trends, and correlations that inform strategic decisions. These insights enable project managers to understand project health, predict potential issues, and optimize resource allocation with unprecedented precision. The key advantage lies in AI's ability to provide real-time, evidence-based recommendations that support timely and accurate decision-making throughout all project phases. This data-driven approach moves project management from intuition-based to intelligence-augmented decision-making, significantly improving project outcomes and success rates.

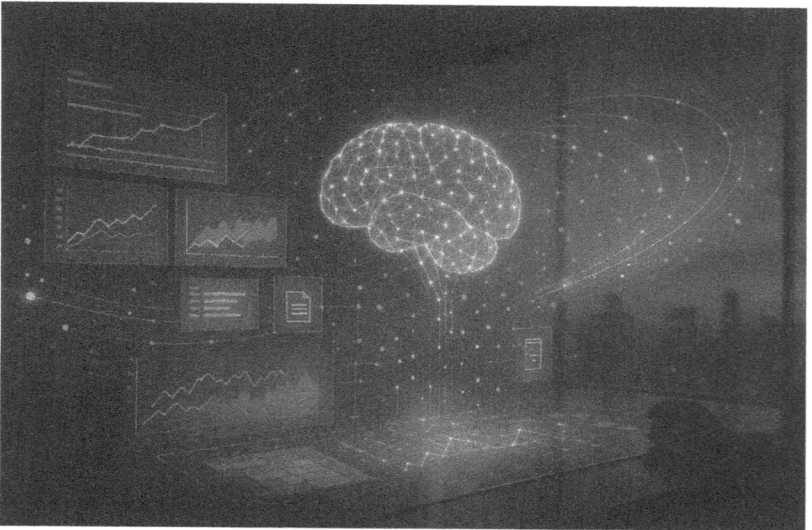

Data-driven risk management represents one of AI's most valuable contributions to project success. AI systems provide early risk identification by continuously monitoring project variables, team performance metrics, and external factors that could impact project delivery. Predictive risk analysis capabilities allow AI to forecast potential issues before they materialize, enabling proactive intervention rather than reactive problem-solving. This predictive capacity supports proactive risk mitigation strategies, where project managers can implement preventive measures based on AI recommendations. The result is a shift from traditional risk management, which often addresses problems after they occur, to predictive risk prevention that maintains project momentum and reduces costly disruptions.

Real-time scenario analysis and forecasting capabilities revolutionize project adaptability and strategic planning. AI-driven simulations can model multiple project scenarios simultaneously, testing various approaches, resource allocations, and timeline adjustments in virtual environments before implementation. This capability enables better decision-making by allowing project managers to evaluate potential outcomes without real-world risks. Dynamic plan adjustments become possible as AI continuously processes new project data and recommends optimization strategies. Real-time forecasting allows project managers to adjust plans dynamically, responding to changing conditions while maintaining project objectives. This adaptability enhances project outcomes and stakeholder satisfaction through improved responsiveness and strategic agility.

Successful AI delegation begins with identifying tasks most suitable for automation based on three key characteristics. Routine and repetitive tasks that follow predictable patterns represent ideal candidates for AI automation, offering immediate efficiency gains and error reduction. Data-intensive processes benefit significantly from AI capabilities, as automated systems can handle large volumes of information more accurately and quickly than human workers. Most importantly, automating these routine activities allows project managers to focus on complex activities requiring human creativity, strategic thinking, and interpersonal skills. The goal is not to replace human capabilities but to optimize the division of labor between human intelligence and artificial intelligence for maximum project effectiveness.

Integrating AI agents into project workflows requires careful alignment with existing processes and human team dynamics. Success depends on aligning AI capabilities with specific workflow requirements, ensuring that automated systems enhance rather than disrupt established processes. This integration must be seamless, matching AI strengths with workflow needs while maintaining productivity. Human-AI collaboration becomes critical, requiring smooth cooperation between team members and AI systems. Project managers must facilitate this collaboration by establishing clear roles, communication protocols, and performance expectations for both human and AI team members. The objective is creating synergistic relationships where AI augments human capabilities rather than creating friction or competition.

Monitoring and evaluating AI agent performance ensures quality, reliability, and alignment with project objectives. Continuous assessment importance cannot be overstated, as ongoing evaluation guarantees that AI outputs meet quality standards and contribute effectively to project success. Regular monitoring ensures AI performance aligns with intended project objectives and outcomes, preventing drift or misalignment that could compromise project delivery. Timely process adjustments become possible through regular evaluation, allowing quick modifications to improve AI-driven processes when performance issues emerge. This systematic approach to AI performance management maintains project quality while maximizing the benefits of AI integration throughout the project lifecycle.

Defining AI's role within the project ecosystem requires recognizing AI as a legitimate stakeholder with specific capabilities and limitations. AI functions as an active participant that influences project outcomes through data analysis, process automation, and decision support capabilities. However, clearly defining AI's responsibilities and limitations ensures effective collaboration and prevents over-reliance on automated systems. Project managers must establish boundaries for AI decision-making authority while leveraging AI strengths in data processing, pattern recognition, and routine task execution. This balanced approach treats AI as a valuable team member with distinct capabilities rather than a replacement for human judgment. Success depends on understanding what AI can and cannot do within your specific project context.

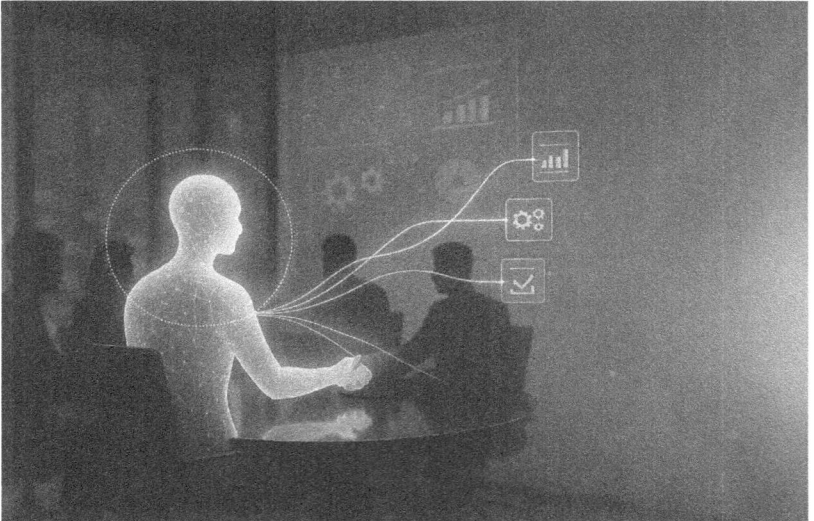

Effective communication and collaboration with AI systems requires developing new interaction skills and interpretation capabilities. Interacting with AI interfaces demands familiarity with user-friendly systems, understanding input requirements, and knowing how to frame queries for optimal AI responses. Project managers must learn to communicate effectively with AI through various interfaces, from simple command systems to complex conversational AI platforms. Interpreting AI insights becomes equally important, as understanding AI-generated data enables informed collaborative decision-making. This interpretation skill involves recognizing AI limitations, understanding confidence levels in AI recommendations, and knowing when human oversight is necessary. These communication skills bridge the gap between human intuition and artificial intelligence capabilities.

Balancing interests between human and AI stakeholders requires careful attention to conflict resolution, value alignment, and shared objectives. Addressing conflicts between human preferences and AI recommendations requires clear protocols for decision-making and dispute resolution. Project managers must establish frameworks for resolving disagreements when human intuition conflicts with AI analysis. Aligning AI with human values ensures that automated systems operate within ethical boundaries and support organizational principles. This alignment requires ongoing monitoring and adjustment of AI parameters to reflect human values and project priorities. Ensuring project objectives means both human and AI stakeholders collaborate effectively toward shared goals, with clear understanding of how each contributes to project success.

Data privacy and security represent fundamental ethical considerations in AI-driven projects that require proactive management and continuous vigilance. The importance of data privacy cannot be understated, as AI systems often process sensitive organizational and personal information that requires protection from unauthorized access and potential breaches. Policy implementation becomes critical, with project managers establishing comprehensive data privacy frameworks that align with regulatory standards like GDPR, HIPAA, or industry-specific requirements. These policies must address data collection, storage, processing, and disposal throughout the AI system lifecycle. Monitoring AI compliance ensures continuous adherence to privacy requirements and maintains robust data security measures, protecting both organizational assets and stakeholder trust.

Bias and fairness in automated decision-making pose significant ethical challenges that project managers must actively address throughout AI implementation. Sources of bias in AI often originate from training data reflecting historical inequities or algorithm design that inadvertently favors certain groups or outcomes. These biases can impact decision-making outcomes negatively, creating unfair advantages or disadvantages for various stakeholders. Project managers have a responsibility to identify and mitigate these biases proactively. Ensuring fairness requires implementing bias detection tools, diverse training datasets, regular algorithm audits, and inclusive development processes. This commitment to fairness protects organizational reputation, ensures equitable outcomes, and maintains stakeholder trust in AI-driven project decisions.

Chapter 2: Evolving Role of the Project Manager

Transparency and accountability for AI actions build stakeholder trust and enable effective governance of automated systems. Clear documentation provides the foundation for AI accountability, requiring comprehensive records of AI decisions, rationale, and performance metrics. This documentation facilitates understanding, enables audit processes, and supports continuous improvement efforts. Building trust through transparent AI explanations helps stakeholders understand how AI systems reach conclusions and make recommendations. Project managers must ensure AI systems can explain their decision-making processes in understandable terms, enabling stakeholders to evaluate and trust AI recommendations. This transparency supports informed decision-making and maintains human oversight over critical project elements while leveraging AI capabilities effectively.

Modern project managers require new technical competencies to effectively lead AI-enabled projects and teams. AI technology understanding forms the foundation, requiring project managers to grasp fundamental AI concepts, capabilities, and limitations to integrate intelligent solutions successfully into their projects. This understanding doesn't require deep technical expertise but sufficient knowledge to make informed decisions about AI implementation and management. Data analytics skills enable project managers to interpret AI-generated insights, understand statistical significance, and make data-driven decisions that optimize project outcomes. Automation tools knowledge helps streamline workflows and improve project efficiency through effective selection, implementation, and management of automated systems that support project objectives.

Chapter 2: Evolving Role of the Project Manager

Digital literacy and adaptability represent core competencies for project managers navigating the AI transformation. Digital literacy empowers project managers to understand, evaluate, and utilize emerging AI technologies effectively, enabling informed decisions about technology adoption and implementation strategies. This literacy includes understanding AI capabilities, recognizing appropriate use cases, and evaluating AI tool effectiveness for specific project needs. Adaptability allows project managers to respond quickly to technological changes and evolving project requirements in dynamic AI-enabled environments. This flexibility includes embracing new tools, adjusting management approaches, and continuously learning as AI capabilities expand. Together, digital literacy and adaptability ensure project managers remain effective leaders as technology continues transforming project management practices.

Developing leadership capabilities for cross-functional, AI-enabled teams requires mastering multiple interconnected competencies. Cross-functional team leadership involves managing diverse teams with varied skills, backgrounds, and expertise levels to achieve shared objectives in AI-augmented environments. Integrating human and AI capabilities requires understanding how to leverage AI tools while maximizing human creativity, judgment, and interpersonal skills for optimal project outcomes. Communication and collaboration skills enable seamless coordination and knowledge sharing across team members and AI systems, ensuring effective information flow and decision-making processes. Change management skills help leaders navigate transitions effectively when adopting new AI technologies and workflows, maintaining team morale and productivity during technological transformation.

In conclusion, this chapter has explored the fundamental transformation of project management in the AI era. The evolving role of project managers encompasses new responsibilities, from managing hybrid teams to leveraging AI insights for strategic decision-making. Embracing ethical practices becomes essential as project managers must ensure responsible AI use, addressing privacy, bias, and accountability concerns proactively. Developing new skillsets represents an ongoing commitment to professional growth, requiring technical competencies, digital literacy, and adaptive leadership capabilities. Success in this AI-enabled future depends on balancing technological capabilities with human judgment, maintaining ethical standards, and continuously evolving your skills to meet emerging challenges and opportunities.

Chapter 3: AI Tool Landscape for PMs

Microsoft Copilot represents a significant advancement in AI-powered project management, offering three core capabilities that directly address common PM challenges. The AI Task Suggestions feature analyzes project data to automatically recommend task prioritization and assignment strategies, reducing manual planning overhead. Timeline Forecasting leverages historical data and current project parameters to predict realistic completion dates, helping managers set achievable deadlines. Progress Insights provides real-time analysis of project health, identifying potential risks before they become critical issues, enabling proactive management rather than reactive problem-solving.

The integration between Microsoft Copilot and Project for the Web creates a unified ecosystem that eliminates data silos and enhances team productivity. Seamless Integration means users work within familiar interfaces while gaining AI capabilities, reducing learning curves and adoption resistance. Real-time Collaboration features ensure all team members access the same updated information simultaneously, preventing miscommunication and duplicated efforts. Data Synchronization maintains consistency across all connected platforms, ensuring project tracking remains accurate. Enhanced Reporting capabilities generate comprehensive insights that support strategic decision-making and stakeholder communication.

The practical benefits of Microsoft Copilot for project managers translate directly into measurable improvements in project outcomes. Improved Scheduling Accuracy results from AI analysis of historical patterns and current constraints, leading to more realistic timelines and reduced project delays. Risk Identification capabilities scan multiple data points to flag potential issues early, allowing teams to implement mitigation strategies before problems escalate. Automated Status Updates eliminate manual reporting tasks, freeing project managers to focus on strategic leadership activities while ensuring stakeholders receive timely, accurate project information.

Notion AI transforms project management workflows through three primary capabilities that address common documentation and organizational challenges. Content Generation assists in creating project briefs, meeting agendas, and status reports quickly, reducing the time spent on administrative tasks. Meeting Notes Synthesis automatically processes discussion points, action items, and decisions from meetings, creating structured summaries that team members can easily reference. Task Prioritization uses intelligent algorithms to suggest which activities should receive immediate attention based on project goals, deadlines, and resource availability.

When comparing Notion AI with other project management assistants, distinct strengths emerge across different platforms. Notion AI excels in Flexible Documentation, offering adaptable templates and structures that can accommodate various project types and organizational needs. ClickUp and Monday.com AI focus on Task Automation, providing robust workflow automation capabilities that streamline repetitive processes and reduce manual intervention. These platforms also offer sophisticated Resource Allocation and Analytics features, using predictive modeling to optimize team capacity and project outcomes through data-driven insights.

AI assistants enhance team productivity through three fundamental improvements that address core collaboration challenges. Improving Transparency occurs when AI systems provide real-time visibility into project status, task progress, and team activities, ensuring all members understand current priorities and deadlines. Reducing Manual Overhead happens through automation of routine activities like status updates, meeting scheduling, and progress tracking, allowing team members to focus on value-added work. Fostering Teamwork results from centralized communication channels and intelligent workflow recommendations that promote better coordination and knowledge sharing among team members.

Chapter 3: AI Tool Landscape for PMs

Jira's AI-powered features specifically target software development and issue tracking challenges that project managers face daily. Automated Issue Tracking uses machine learning to categorize, prioritize, and route issues based on content analysis and historical patterns, reducing manual triage time and improving response accuracy. Backlog Prioritization analyzes multiple factors including business value, technical dependencies, and resource availability to suggest optimal task sequencing. Project Risk Prediction examines velocity trends, issue patterns, and team capacity to identify potential bottlenecks and delivery risks before they impact project timelines.

Asana's AI enhancements focus on optimizing resource management and deadline forecasting to improve project execution. Task Assignment Optimization analyzes team member skills, current workload, and availability to recommend optimal task distribution, ensuring work is assigned to the most suitable resources. Deadline Forecasting uses historical completion data and current project parameters to predict realistic delivery dates, helping teams set achievable milestones. Actionable Insights identify workflow bottlenecks, resource constraints, and performance trends, providing project managers with data-driven recommendations for process improvements and team optimization.

Trello's approach to AI integration emphasizes automation and extensibility through strategic partnerships and built-in features. AI-Powered Automation, primarily through Butler, enables users to create custom rules that automatically move cards, assign team members, set due dates, and trigger notifications based on predefined conditions. This reduces manual board management and ensures consistent workflow execution. Third-Party Integrations expand Trello's capabilities by connecting with external AI tools and services, allowing teams to incorporate specialized functionality like sentiment analysis, predictive analytics, and advanced reporting into their project workflows.

Microsoft Power Platform enables organizations to create sophisticated AI automations without requiring extensive programming expertise. Tailored AI Workflows allow teams to design custom solutions that address specific organizational needs and processes, going beyond generic project management features. Data Integration capabilities connect multiple enterprise systems, creating unified data sources that power more accurate AI insights and predictions. The platform's No Extensive Coding Required approach democratizes AI development, enabling business users to create automation solutions through visual interfaces and pre-built connectors, reducing dependency on technical resources.

Zapier AI serves as a powerful integration platform that connects disparate project management tools through intelligent automation. App Integration capabilities span thousands of applications, enabling seamless data flow between project management platforms, communication tools, file storage systems, and business applications. Automation of Notifications ensures team members receive timely updates across all relevant channels, reducing information gaps and improving response times. Data Updates and Task Triggers create intelligent workflows that automatically update project information, create tasks, and initiate processes based on specific conditions or events across connected applications.

Custom AI applications address specific project management needs through targeted solutions that complement existing tools. Automated Risk Assessment systems analyze project data from multiple sources to create comprehensive risk dashboards, providing early warning indicators and suggested mitigation strategies. Resource Allocation Models use optimization algorithms to recommend optimal team assignments, budget distribution, and timeline adjustments based on current constraints and historical performance. Intelligent Project Reporting generates customized reports that adapt to stakeholder needs, automatically highlighting critical issues, milestone progress, and performance metrics relevant to different audiences.

Understanding the distinction between embedded and standalone AI solutions is crucial for making informed technology decisions. Embedded AI Tools integrate directly within existing project management platforms, appearing as native features that enhance current functionality without requiring separate interfaces or data migration. These solutions typically offer seamless user experiences but may have limited customization options. Standalone AI Solutions operate as independent applications that connect to other systems through APIs or data connectors, providing specialized capabilities and greater flexibility but potentially requiring additional integration effort and management overhead.

Each AI approach offers distinct advantages and faces specific limitations that influence implementation decisions. Embedded AI Benefits include seamless user experience, tight system integration, and minimal latency, making adoption easier for teams already comfortable with existing tools. However, Embedded AI Limitations involve restricted customization options due to platform constraints and fixed feature sets. Standalone AI Benefits provide greater flexibility, specialized capabilities, and easier customization for unique organizational needs. Standalone AI Challenges include integration complexity, potential data silos, and additional management resources required for maintenance and updates.

Selecting the optimal AI strategy requires careful consideration of multiple organizational factors and project requirements. Embedded AI Integration works best for teams seeking immediate productivity gains within familiar tools, offering lower risk and faster adoption timelines. Standalone AI Solutions suit organizations with specific requirements, technical resources for integration, and needs for specialized capabilities not available in embedded options. Key Factors Influencing AI Choice include team size and technical expertise, project complexity and customization requirements, existing tool investments, budget constraints, and long-term scalability needs.

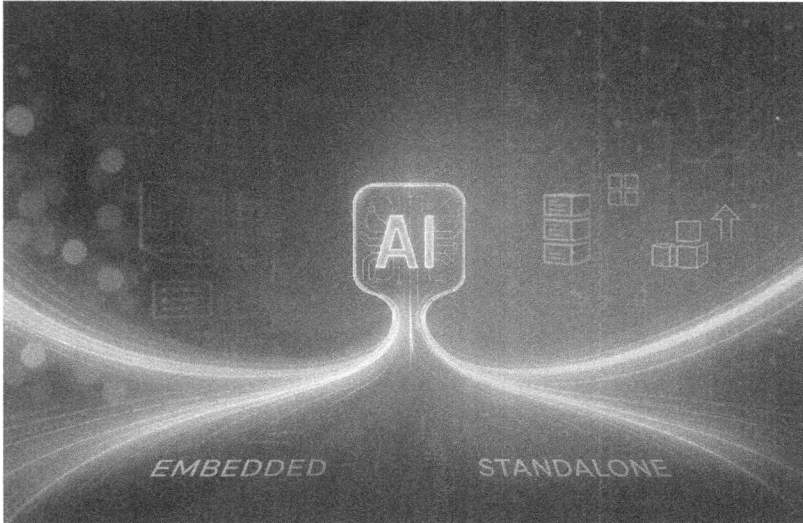

AI technologies represent a transformative opportunity for project management professionals to enhance their effectiveness and project outcomes. AI Enhances Efficiency by automating routine tasks, optimizing resource allocation, and providing intelligent insights that accelerate decision-making processes. Improved Collaboration results from better communication tools, shared visibility, and automated coordination mechanisms that keep teams aligned and productive. Informed Decision-Making becomes possible through data-driven insights, predictive analytics, and real-time monitoring capabilities that support strategic project leadership. The future of project management lies in effectively leveraging these AI capabilities while maintaining focus on human leadership and creativity.

Chapter 4: AI in Project Initiation

AI is transforming project management from the very beginning. Four key applications define this transformation: automated data gathering streamlines information collection from multiple sources, improving efficiency and accuracy. Business case generation becomes more robust as AI analyzes historical data and predicts outcomes. Stakeholder identification leverages communication analysis to map key participants and their interests. Early risk assessment uses machine learning to identify potential issues before they become problems. These applications work together to create a comprehensive foundation for successful project initiation, enabling project managers to make more informed decisions with greater confidence.

AI adoption in early project stages presents both significant benefits and notable challenges. The primary benefit is accelerated analysis with enhanced accuracy, enabling faster and more precise decision-making. However, organizations face substantial challenges including data quality issues and complex system integration requirements. Successfully implementing AI requires effective change management strategies and building stakeholder trust in automated outputs. Organizations must address these challenges systematically, ensuring proper data governance, seamless technology integration, and comprehensive training programs. The key to success lies in balancing technological capabilities with human expertise and maintaining transparency in AI-driven processes.

AI revolutionizes business case development through three critical processes. Automatic data gathering enables AI systems to efficiently collect and aggregate diverse datasets from multiple organizational sources, creating comprehensive analytical foundations. Advanced trend identification algorithms detect significant patterns within large datasets that human analysts might miss. Finally, AI synthesizes these insights to build evidence-based business cases rapidly, combining quantitative analysis with predictive modeling. This automated approach reduces the time required for business case development from weeks to days while improving accuracy and completeness. The result is more compelling, data-driven justifications for project investments.

AI-driven scenario modeling enhances project planning through sophisticated simulation capabilities. AI models can simulate multiple project scenarios simultaneously, exploring various potential outcomes and identifying possible challenges before they occur. Outcome forecasting uses historical data and trend analysis to predict project results with greater accuracy, improving planning precision and stakeholder readiness. Stress-testing assumptions becomes more rigorous as AI evaluates different variables under various conditions, helping teams understand potential vulnerabilities. This comprehensive approach to scenario planning enables project teams to develop more robust strategies and contingency plans, ultimately increasing project success rates.

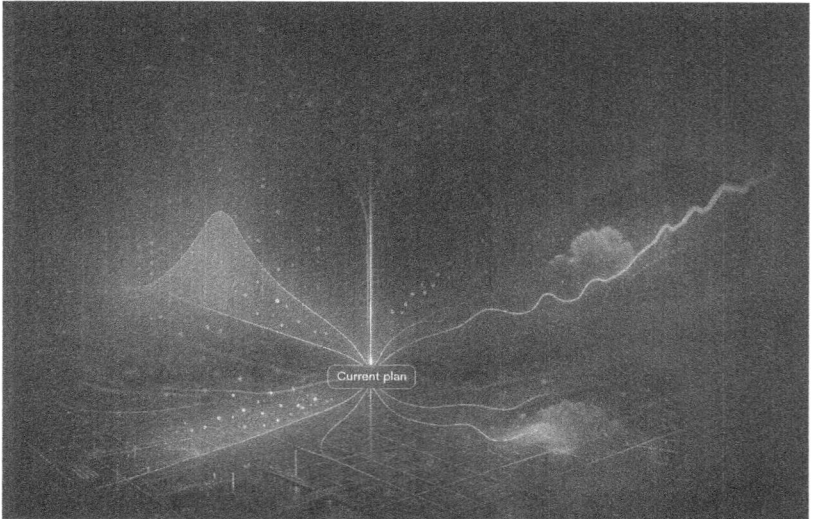

AI transforms decision-making by providing deeper analytical insights throughout the project initiation process. AI systems excel at identifying key drivers by analyzing complex project data to reveal critical factors that influence outcomes, often uncovering relationships that weren't previously apparent. Risk detection capabilities allow AI to identify potential issues early in the project lifecycle, enabling proactive mitigation strategies. Most importantly, these AI-powered insights empower stakeholders to make well-informed decisions with greater confidence, reducing uncertainty and improving project outcomes. The combination of enhanced analysis and early warning systems creates a more strategic approach to project initiation.

Natural Language Processing transforms stakeholder identification by analyzing textual communications across multiple channels. NLP algorithms systematically examine emails, documents, meeting transcripts, and social media interactions to identify key stakeholder information and interests. This automated analysis reveals stakeholder networks and influence patterns that might be missed through traditional manual methods. The comprehensive data sources scanned by NLP provide a complete picture of stakeholder engagement and concerns. Project managers can then create accurate stakeholder maps based on objective data rather than assumptions, ensuring no critical stakeholders are overlooked during project planning and improving overall stakeholder management effectiveness.

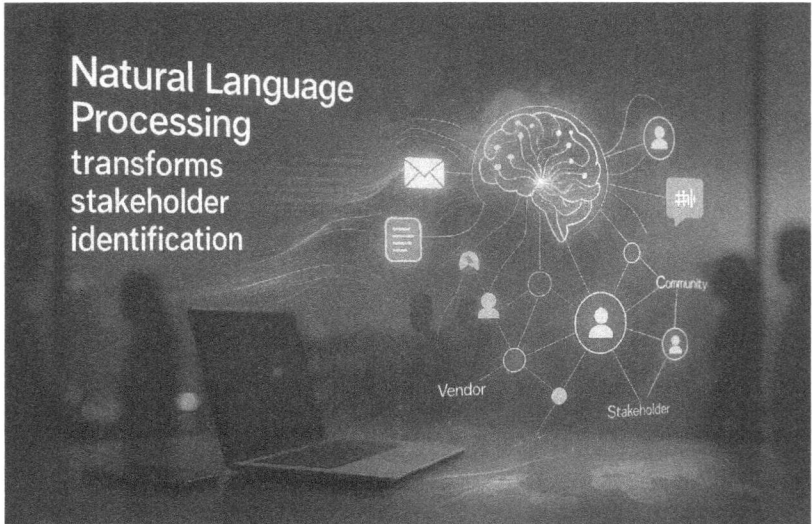

Natural Language Processing transforms stakeholder identification

AI-powered communication analysis provides unprecedented insights into stakeholder sentiment and engagement levels. Tone analysis examines communication patterns to understand stakeholder emotions and attitudes toward project initiatives, identifying potential supporters and resisters. Frequency monitoring tracks communication levels to determine engagement intensity and interest in project outcomes. These analyses reveal crucial engagement and sentiment insights that help project managers understand stakeholder dynamics and potential challenges. By combining tone analysis with frequency patterns, teams can predict stakeholder behavior and develop targeted engagement strategies. This proactive approach to stakeholder management significantly improves project buy-in and reduces resistance.

AI visualization tools create powerful stakeholder influence networks that transform project management approaches. AI-driven visual mapping generates clear, comprehensive representations of stakeholder relationships and influence hierarchies, making complex networks understandable at a glance. Understanding stakeholder influence becomes more precise as visual maps identify key decision-makers and their interconnections within stakeholder groups. These visualizations enable targeted communication strategies, allowing project managers to tailor their approaches to specific stakeholder groups and influence pathways. The result is more effective stakeholder engagement, improved communication efficiency, and better project outcomes through strategic relationship management.

Predictive analytics revolutionizes early risk detection by identifying potential problems before they manifest. AI-driven pattern recognition analyzes historical project data and current conditions to detect early warning signs of potential risks. These sophisticated models can identify subtle patterns and correlations that human analysts might miss, providing comprehensive risk visibility. Proactive risk mitigation becomes possible as early identification enables teams to address issues before they escalate into major problems. This predictive approach transforms risk management from reactive to proactive, significantly improving project success rates. Teams can allocate resources more effectively and develop targeted mitigation strategies based on AI-generated risk insights.

AI transforms opportunity recognition by analyzing vast datasets to identify strategic possibilities that align with organizational goals. Data analysis capabilities enable AI platforms to detect promising opportunities within complex business environments, revealing potential value creation that might otherwise be missed. Prioritization becomes more strategic as AI assesses potential impact and feasibility, helping organizations focus on the most valuable opportunities. This systematic approach to opportunity identification ensures resources are directed toward initiatives with the highest probability of success and greatest potential return. AI-driven opportunity recognition enables organizations to stay competitive by identifying emerging trends and market possibilities early.

Successfully integrating AI findings into project planning requires systematic approaches across three key areas. AI-driven risk identification provides early warning systems that enable proactive mitigation strategies, reducing project failures and cost overruns. Opportunity recognition capabilities uncover hidden value creation possibilities that optimize project outcomes and resource allocation decisions. Systematic planning integration ensures AI findings are properly incorporated into comprehensive project frameworks. This integration approach transforms project planning from intuition-based to data-driven decision-making. The result is more robust project plans that account for both risks and opportunities, leading to improved success rates and better resource utilization.

AI enhances technical feasibility assessment through advanced analytical capabilities that process complex technical requirements efficiently. Data analysis capabilities enable AI to examine technical constraints, system requirements, and implementation challenges systematically. AI can evaluate technical specifications against available resources and capabilities, identifying potential compatibility issues and resource gaps. Predicting implementation challenges becomes more accurate as AI analyzes similar past projects and identifies common technical obstacles. This proactive approach to technical assessment reduces project delays and cost overruns by addressing potential issues during the planning phase. Technical teams can make more informed decisions about technology choices and implementation strategies.

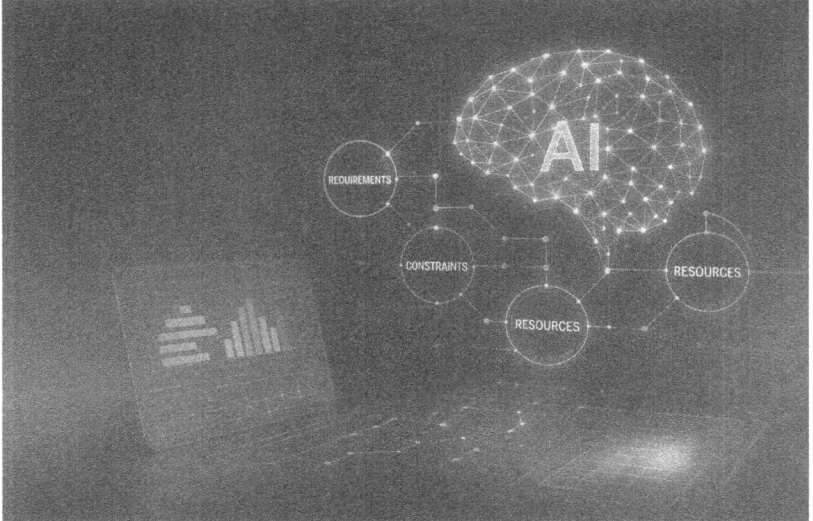

AI tools significantly enhance market and financial feasibility analysis through sophisticated forecasting capabilities. Market trend analysis uses AI to examine consumer behavior patterns, competitive landscapes, and demand forecasting with greater accuracy than traditional methods. Financial model evaluation becomes more precise as AI analyzes cost structures, revenue projections, and profitability scenarios using historical data and predictive modeling. Enhanced feasibility accuracy results from combining market intelligence with financial forecasting, providing comprehensive business case validation. This integrated approach ensures projects are both technically viable and commercially sound, reducing investment risks and improving decision-making quality for stakeholders and investors.

Continuous feasibility monitoring represents a paradigm shift from static to dynamic project assessment throughout initiation phases. Ongoing feasibility monitoring uses AI to track multiple metrics simultaneously, providing real-time insights into changing project conditions and market dynamics. Automated alert systems notify teams immediately when critical parameters shift, enabling rapid response to changing circumstances. This supports agile decision-making by providing updated feasibility data continuously rather than at fixed intervals. Teams can adapt their strategies quickly based on current information, maintaining project viability even as conditions change. This dynamic approach significantly improves project success rates by ensuring decisions remain relevant and appropriate.

This chapter demonstrates how AI integration transforms project initiation through three fundamental improvements. Enhanced analysis capabilities provide deeper insights and more comprehensive data-driven analysis during critical early phases, enabling better decision-making. Improved stakeholder engagement results from AI tools that facilitate more effective communication and collaboration among all project participants from the very beginning. Proactive risk management becomes possible as AI identifies and helps mitigate potential issues early, dramatically improving project success rates. Organizations implementing these AI capabilities gain significant competitive advantages through more strategic, data-driven project initiation processes that deliver superior outcomes.

Chapter 5: AI in Planning

AI integration in project planning represents a fundamental shift from reactive to predictive management approaches. Machine learning algorithms continuously analyze vast datasets from historical projects, identifying patterns that humans might miss and optimizing timelines with remarkable precision. Natural language processing capabilities enable AI systems to interpret complex project documentation, automatically extract requirements, and facilitate seamless communication between stakeholders. Advanced analytics provide real-time insights that enhance decision-making processes, allowing project managers to allocate resources more effectively and anticipate potential challenges before they become critical issues.

While AI-driven planning offers substantial benefits, organizations must navigate several key challenges to realize its full potential. Enhanced decision-making capabilities emerge from AI's ability to rapidly process complex data and provide actionable insights based on comprehensive analysis. Risk reduction becomes achievable through early identification and mitigation of potential project threats. However, data quality remains a critical challenge, as inaccurate or incomplete information can significantly compromise AI effectiveness and lead to suboptimal outcomes. Successfully adopting AI requires careful change management and ensuring AI models remain transparent and interpretable to stakeholders.

The evolution of AI in project management is accelerating rapidly, with several emerging trends reshaping the landscape. Explainable AI is becoming increasingly important, as organizations demand transparency in AI decision-making processes to build trust and accountability. Integration with digital twins enables sophisticated simulation and analysis of projects in virtual environments, allowing for comprehensive planning scenarios. Real-time data analytics provide instantaneous insights that facilitate agile and adaptive project management approaches. Increased automation is streamlining repetitive tasks, freeing project managers to focus on strategic activities that require human expertise and creativity.

AI revolutionizes task decomposition by systematically analyzing project objectives and breaking them into manageable components. The AI system examines project goals comprehensively, understanding scope, requirements, and interdependencies with remarkable thoroughness and accuracy. Complex objectives are automatically divided into smaller, manageable tasks that streamline project execution and improve workflow efficiency. This automation significantly minimizes human workload while improving accuracy in task organization and sequencing. Project managers can now focus on strategic oversight rather than manual task breakdown, resulting in more consistent and comprehensive work breakdown structures across all project types.

AI enhances project scope definition through sophisticated data analysis and pattern recognition capabilities. By analyzing historical project data, AI systems identify recurring patterns and assist in defining accurate project boundaries and deliverables. The technology ensures project scope aligns closely with defined goals and organizational objectives, promoting higher success rates and stakeholder satisfaction. AI helps minimize scope creep, a common project management challenge, by providing clarity and consistency in scope definition from project inception. This systematic approach to scope definition reduces ambiguity and creates a solid foundation for all subsequent project planning activities.

Real-world implementations of AI-generated work breakdown structures demonstrate significant improvements across multiple dimensions of project management. Planning speed increases dramatically as AI automates the traditionally time-intensive process of task decomposition and sequencing. Enhanced clarity emerges through AI's ability to create well-structured, logical work breakdowns that improve understanding among project team members and stakeholders. AI-generated structures facilitate superior communication and alignment between diverse stakeholder groups and project teams, reducing misunderstandings and conflicts. These improvements translate directly into reduced project initiation time and more accurate resource planning from the earliest stages of project development.

Natural language processing transforms schedule creation by enabling AI systems to interpret and understand project requirements expressed in everyday language. NLP algorithms analyze natural language prompts with sophisticated understanding, extracting key requirements, timelines, and dependencies from conversational descriptions. This capability eliminates the need for complex technical input formats, making project planning more accessible to stakeholders across various skill levels. Automated schedule generation occurs with high accuracy, dramatically reducing manual effort and potential human errors. The system can process complex, multi-faceted project descriptions and convert them into comprehensive, actionable project schedules ready for implementation.

AI-driven visualization capabilities enable dynamic, real-time updates of project timelines that accurately reflect current progress and changes. These systems provide immediate timeline adjustments as project conditions evolve, ensuring stakeholders always have access to current information. Improved project tracking becomes possible through enhanced accuracy in monitoring milestones, deadlines, and critical path activities. AI-driven visualizations facilitate clearer communication among stakeholders through intuitive, easily understood timeline displays that reduce complexity and improve comprehension. This enhanced communication capability leads to better decision-making and more effective project coordination across all organizational levels.

Chapter 5: AI in Planning

The efficiency gains from AI-generated Gantt charts are substantial and measurable across multiple project dimensions. Automated chart creation significantly reduces planning time while simultaneously minimizing errors that commonly occur in manual schedule development. Project managers can redirect their time from tedious scheduling tasks to more strategic activities like risk management, stakeholder engagement, and team leadership. This shift in focus allows managers to dedicate more attention to project execution and proactive risk mitigation strategies. The reduction in planning errors creates a more reliable foundation for project success and improved stakeholder confidence.

AI algorithms revolutionize resource matching by analyzing multiple complex factors simultaneously to optimize assignments. The system evaluates skills, availability, workload capacity, and project priorities to assign the most suitable resources to specific tasks. This comprehensive analysis ensures optimal resource utilization while maintaining quality standards and meeting project deadlines. Efficient scheduling becomes possible through AI's ability to consider multiple constraints and variables that would be overwhelming for manual planning. The result is improved resource utilization rates, reduced conflicts, and enhanced project delivery capabilities across the entire organization.

Continuous workload analysis enables AI systems to monitor resource allocation in real-time, identifying imbalances before they impact project performance. The technology prevents both resource overuse and underuse, ensuring balanced team efforts and efficient task distribution across all project phases. This balanced approach leads to maximized productivity through optimal workload distribution that considers individual capabilities and organizational constraints. Teams experience improved morale and performance when workloads are appropriately balanced, leading to higher quality deliverables and reduced burnout risks. The system continuously adapts to changing project conditions, maintaining optimal resource allocation throughout the project lifecycle.

Dynamic resource adjustments represent a significant advancement in project management flexibility and responsiveness. AI enables real-time adaptation of resource allocation to meet evolving project demands and capitalize on emerging opportunities. Scenario planning capabilities help project managers anticipate potential changes and prepare appropriate responses, optimizing project outcomes under various conditions. This proactive approach allows organizations to maintain project momentum even when facing unexpected challenges or opportunities. The combination of dynamic adjustments and comprehensive scenario analysis creates more resilient project management capabilities that can adapt to changing business environments.

Predictive analytics transforms cost estimation through sophisticated analysis of historical project data and pattern recognition. AI systems identify cost patterns from extensive historical datasets, improving forecasting accuracy and reducing financial surprises. This enhanced accuracy in cost prediction significantly reduces risks associated with budget overruns and unexpected financial challenges. Budget allocation optimization becomes possible through AI-driven forecasts that enable teams to allocate resources more strategically and efficiently. Organizations can make more informed financial decisions, improving overall project profitability and reducing financial risks associated with project delivery.

Machine learning revolutionizes timeline forecasting by identifying subtle patterns in project data that indicate potential schedule variations. Pattern recognition capabilities enable the system to learn from past project experiences, continuously improving timeline accuracy and reliability. Delay prediction becomes proactive rather than reactive, as models analyze current progress indicators to forecast potential scheduling challenges. This predictive capability allows project managers to implement corrective measures before delays become critical, maintaining project momentum and stakeholder confidence. The system's ability to learn and adapt ensures that forecasting accuracy improves over time with each completed project.

AI-driven risk management provides early detection capabilities that identify potential project threats before they materialize into significant problems. The system analyzes multiple data points simultaneously, recognizing risk patterns that might be invisible to human analysis. Proactive risk mitigation becomes possible through AI-generated insights that suggest specific measures to reduce identified risks and enhance project success probability. This forward-looking approach transforms risk management from reactive problem-solving to strategic prevention, significantly improving project outcomes. Organizations benefit from reduced project failures, lower costs associated with risk events, and improved stakeholder confidence in project delivery capabilities.

Digital twins create precise virtual replicas of physical projects, enabling comprehensive visualization and analysis without impacting actual project execution. These virtual environments allow teams to test changes and scenarios safely, exploring various options without risking real-world consequences or delays. Risk-free testing capabilities enable project managers to evaluate multiple approaches and select optimal strategies before implementation. Digital twins enhance project planning by enabling real-time analysis and informed decision-making based on comprehensive simulation results. This technology represents a significant advancement in project planning sophistication and strategic capability.

AI-powered digital twins simulate complex real-world processes with remarkable accuracy, enabling sophisticated scenario creation and analysis capabilities. The system can generate and analyze multiple scenarios simultaneously, predicting potential outcomes under various conditions and constraints. This comprehensive analysis supports strategic and informed decision-making in complex project environments where multiple variables interact dynamically. Project managers gain unprecedented insight into project behavior under different conditions, enabling more robust planning and contingency preparation. The combination of AI processing power and digital twin technology creates powerful planning tools previously unavailable to project managers.

Digital twins significantly enhance project adaptability and resilience through accurate forecasting of change impacts on project outcomes. This capability enables proactive decision-making that anticipates and prepares for various scenarios and potential disruptions. Enhanced project resilience emerges from the ability to prepare for and respond effectively to uncertainties, disruptions, and changing requirements. Organizations can maintain project momentum even when facing significant challenges or market changes. The technology creates more robust project management capabilities that adapt dynamically to changing conditions while maintaining focus on successful project delivery.

The conclusion emphasizes four critical benefits of AI adoption in project management. Automation of complex tasks reduces manual effort while significantly increasing overall productivity and accuracy. Enhanced accuracy in planning and project management minimizes errors and associated risks, leading to more predictable project outcomes. Better decision-making capabilities emerge from AI's ability to provide insightful data and predictive analytics that inform strategic choices. Embracing AI technologies creates a clear pathway to project success through more efficient execution methods and dramatically higher success rates across all project types and organizational contexts.

Chapter 6: AI in Execution and Monitoring

AI integration fundamentally transforms project workflows through three core capabilities. First, automation of repetitive tasks significantly reduces manual effort while increasing overall workflow efficiency across all project phases. This includes automated data entry, status updates, and routine administrative functions that traditionally consume valuable team time. Second, AI excels at complex data analysis, processing large and intricate datasets to uncover patterns and insights that would be impossible for humans to identify manually. Finally, AI delivers actionable insights that directly guide project decisions, leading to improved outcomes and enhanced efficiency. These three pillars work together to create a comprehensive project management ecosystem that supports better decision-making and streamlined operations.

Chapter 6: AI in Execution and Monitoring

Real-time AI-driven execution and monitoring delivers three critical benefits for project success. Timely project visibility provides immediate insights into project status, enabling better decision-making by eliminating information delays and ensuring stakeholders always have current data. Faster issue response becomes possible as AI enables rapid detection and response to project issues, significantly reducing delays and associated risks before they escalate into major problems. Improved resource utilization ensures efficient use of both manpower and materials through AI optimization algorithms that continuously analyze workload distribution and resource availability. These benefits combine to create a more responsive, efficient, and successful project environment where issues are caught early and resources are maximized for optimal project outcomes.

Successful AI implementation requires careful consideration of four key challenges. Data quality importance cannot be overstated, as high-quality, consistent data is absolutely essential for effective AI implementation and accurate decision-making processes. Poor data quality leads to unreliable AI insights and flawed recommendations. Managing organizational change presents another critical challenge, requiring effective change management strategies to support smooth transitions within organizations as teams adapt to new AI-powered workflows. System integration challenges demand seamless compatibility and technical alignment between AI tools and existing project management systems. Finally, user adoption and ethics remain critical, ensuring team members embrace AI tools while maintaining transparency and ethical AI use throughout the implementation process for long-term organizational success.

AI revolutionizes task assignment through comprehensive analysis of team capabilities and current workloads. Assessment of team skills involves AI evaluating individual capabilities, technical competencies, and expertise areas to identify strengths and specializations within the team. This creates detailed skill profiles that inform optimal task matching. Workload analysis continuously monitors current task distribution to ensure assignments are evenly balanced among team members, preventing burnout and bottlenecks. The system tracks capacity, availability, and work-in-progress items for each team member. Task assignment optimization represents the culmination of this analysis, where AI recommends the best fit for each task based on both skills alignment and current workload capacity, maximizing overall team efficiency and project success while ensuring sustainable work distribution.

Dynamic assignment based on project priorities enables teams to respond effectively to changing project demands. Real-time task allocation ensures assignments adjust instantly based on shifting project priorities, allowing teams to optimize focus on the most critical work items as business needs evolve. This flexibility prevents teams from working on outdated priorities. Focus on high-impact activities helps teams prioritize tasks that deliver the greatest value and impact, ensuring alignment with evolving project goals and business objectives. Resources are directed toward work that matters most. Flexible response to demands enables teams to adapt quickly to new requirements, unexpected challenges, and project changes without losing momentum or efficiency. This adaptability is crucial in today's fast-paced business environment where priorities can shift rapidly.

AI-driven task assignment creates significant positive impacts on both productivity and employee satisfaction. Optimized task assignments effectively reduce workflow bottlenecks and prevent employee burnout by ensuring work is distributed according to capacity and capability rather than arbitrary assignment. This strategic approach eliminates overloading high performers while underutilizing others. Increased productivity naturally results as reducing bottlenecks leads to smoother operations and higher overall team productivity, with work flowing more efficiently through the project pipeline. Enhanced job satisfaction emerges from balanced workloads that improve employee satisfaction and promote a positive work environment. When team members receive assignments that match their skills and capacity, they feel more engaged, valued, and capable of producing high-quality work, leading to better retention and team morale.

Real-time data collection forms the foundation of automated status reporting through seamless AI integration with project management tools. AI integration connects effortlessly with multiple project tools including task managers, time tracking systems, and collaboration platforms to automate continuous data collection without requiring manual intervention from team members. This eliminates data silos and ensures comprehensive information gathering. Continuous data gathering ensures project tools provide ongoing, real-time data updates without requiring manual input from users, creating a constantly updated picture of project status. Up-to-date insights result from this real-time data collection, ensuring project insights are always current and accurate, enabling managers to make informed decisions based on the latest information rather than outdated reports that may misrepresent actual project status.

Automated generation of progress reports streamlines the reporting process through three key stages. Data collection begins with AI gathering accurate and relevant information from various integrated sources to ensure report precision and completeness. This includes pulling data from multiple project tools, databases, and systems to create a comprehensive view of project status. Report generation follows, where AI automatically creates customized progress reports that clearly highlight key metrics, trends, and critical information tailored to different stakeholder needs. These reports maintain consistency while adapting content for different audiences. Stakeholder clarity is achieved through generated reports that present information clearly and concisely, supporting stakeholder decision-making and progress tracking by providing actionable insights rather than raw data dumps, enabling quick comprehension and informed decision-making.

Automation significantly reduces manual effort and human error in reporting processes, delivering three primary advantages. Time savings through automation dramatically reduces the time required to complete repetitive reporting tasks, freeing up valuable team resources for more strategic activities. What previously took hours of manual compilation can now be accomplished in minutes. Reduction of human errors occurs as automated systems minimize common mistakes caused by manual data entry, transcription errors, and calculation mistakes in reports. Consistent formatting and accurate data processing become standard. Improved reporting reliability results from automation enhancing both accuracy and reliability of reporting outcomes, ensuring stakeholders receive trustworthy information for decision-making. Automated reports maintain consistent quality and format, reducing variability and improving overall communication effectiveness across the organization.

Chapter 6: AI in Execution and Monitoring

Automated documentation of meeting discussions leverages AI capabilities to enhance meeting productivity and record-keeping. AI transcription technology automatically captures and transcribes spoken content during meetings, providing accurate documentation without requiring dedicated note-takers. This ensures complete conversation capture including nuances, decisions, and action items that might otherwise be missed. Advanced speech recognition technology handles multiple speakers, accents, and technical terminology. Accurate summaries are generated by AI to create precise and clear meeting summaries that enhance understanding and support comprehensive record-keeping. These summaries extract key decisions, action items, and important discussion points, presenting them in organized formats that team members can quickly reference. This eliminates the need for manual note-taking and ensures consistent, professional documentation of all meeting content.

Intelligent reminders and follow-ups ensure meeting outcomes translate into actionable results. AI-powered reminders deliver timely notifications to team members, ensuring important tasks and commitments made during meetings are not forgotten post-meetings. The system tracks deadlines, sends appropriate reminders based on urgency and individual preferences, and escalates as needed. This proactive approach prevents items from falling through cracks. Action item tracking involves AI monitoring action items and associated deadlines, helping teams maintain momentum and meet project goals efficiently. The system provides visibility into progress, identifies potential delays, and ensures accountability. Team members receive regular updates on their commitments and can easily track what needs attention, creating a culture of follow-through and reliable execution of meeting decisions and commitments.

Chapter 6: AI in Execution and Monitoring

AI-powered meeting management significantly boosts team alignment and accountability through improved processes. Clear record keeping enables transparent and accessible documentation of team activities, enhancing clarity and trust among team members. Everyone has access to the same information, reducing misunderstandings and ensuring consistent communication. Detailed meeting records provide reference points for future discussions and decision-making. Follow-up mechanisms through automated reminders and tracking help teams stay on schedule and complete tasks efficiently, ensuring meeting decisions translate into concrete actions. This systematic approach prevents important items from being overlooked. Enhanced team accountability emerges as stronger coordination supported by AI fosters responsibility and alignment among team members. When everyone can see commitments, deadlines, and progress, accountability naturally increases, leading to better team performance and project outcomes.

Machine learning models revolutionize project timeline forecasting through sophisticated data analysis capabilities. Data-driven predictions utilize both historical project data and current project information to accurately predict completion dates and resource allocation requirements. These models learn from past project patterns, identifying factors that typically influence timeline success or delays. The system considers multiple variables including team performance, resource availability, and external dependencies. Improved project planning results as machine learning enhances managers' ability to plan projects by providing precise forecasts of timelines and resource needs. Rather than relying on gut instinct or simple estimates, managers receive data-backed projections that account for complexity, risk factors, and team capabilities. This leads to more realistic project schedules, better resource planning, and improved stakeholder expectations management.

Early detection of schedule or scope deviation enables proactive project management through predictive analytics. The predictive analytics role involves analyzing project patterns and trends to detect early risks of schedule delays and scope changes before they become critical issues. Advanced algorithms identify warning signs such as velocity decreases, resource constraints, or scope creep indicators. This early warning system examines multiple data points to provide comprehensive risk assessment. Timely intervention becomes possible as early risk detection enables project managers to implement corrective measures before problems escalate. Rather than responding to crises, teams can address issues while they're still manageable, preventing minor setbacks from becoming major delays. This proactive approach significantly improves project success rates and reduces the stress associated with last-minute problem-solving and crisis management.

Automated alerts and recommended corrective actions provide comprehensive risk management support for project teams. Automated risk notifications ensure AI systems generate real-time alerts to identify potential risks quickly and accurately, providing immediate awareness of emerging issues. These alerts are intelligently prioritized based on severity and impact, ensuring teams focus on the most critical concerns first. Actionable corrective steps are suggested by the system, providing specific, practical recommendations to mitigate identified risks effectively and promptly. Rather than simply flagging problems, the AI offers concrete solutions and action plans. Enhanced decision-making results from automated alerts improving responsiveness and supporting better decision-making in critical situations. Teams receive not just warnings, but also the information and recommendations needed to address issues confidently and effectively, leading to faster resolution and better project outcomes.

Real-time identification of emerging risks provides continuous protection for project success through advanced monitoring capabilities. AI data monitoring involves continuous analysis of diverse data sources to identify potential risks in real time, examining patterns across multiple project dimensions including schedule, budget, resources, and quality metrics. The system processes information from various tools and sources to provide comprehensive risk visibility. Early risk warnings are delivered through timely alerts to project teams, enabling proactive risk management and mitigation before issues escalate into serious problems. These warnings are intelligently filtered and prioritized to avoid alert fatigue while ensuring critical risks receive immediate attention. This proactive approach transforms risk management from reactive crisis response to preventive project protection, significantly improving project success rates and team confidence.

AI-driven risk assessment strategies provide comprehensive evaluation and management of project uncertainties. Quantifying risk probabilities involves AI algorithms analyzing historical and current data to accurately quantify the likelihood of various risks occurring, providing statistical confidence levels for different risk scenarios. This quantitative approach moves beyond subjective risk assessment to data-driven probability analysis. Assessing potential impacts utilizes advanced models to evaluate the potential consequences of identified risks, informing better decision-making by understanding not just the likelihood but also the severity of potential issues. Prioritized risk management emerges as AI supports ranking and prioritizing risks to optimize management strategies and resource allocation. Teams can focus their limited time and resources on the risks that pose the greatest threat to project success, creating more efficient and effective risk management processes.

Integrating risk insights into decision-making ensures project management becomes truly data-driven and strategic. AI-driven risk data generation provides comprehensive risk information that supports informed decision-making and enhances strategic planning capabilities. Teams receive actionable intelligence rather than raw data, enabling confident decision-making based on clear risk profiles and recommendations. Strategy adjustments become possible as risk insights enable timely modifications to address emerging uncertainties effectively. Projects can adapt their approach, timeline, or resource allocation based on changing risk landscapes. Resource allocation benefits from integrated risk data that guides optimal distribution of resources to mitigate potential risks. Rather than spreading resources evenly, teams can concentrate efforts where they'll have the greatest impact on risk reduction and project success, creating more efficient and effective project management approaches that proactively address challenges.

Chapter 6: AI in Execution and Monitoring

This chapter has demonstrated how AI empowers modern project management through intelligent automation and advanced monitoring capabilities. AI enables streamlined project execution by automating routine tasks while providing proactive monitoring insights that keep projects on track. Enhanced team collaboration emerges through advanced AI strategies that foster better communication and coordination across all project phases, from planning through delivery. Risk mitigation becomes more effective as AI-driven insights help identify and reduce project risks before they escalate into serious problems. By implementing these AI-powered strategies, teams can achieve higher success rates, improved efficiency, and better project outcomes. The future of project management lies in leveraging these intelligent systems to augment human capabilities and create more resilient, responsive, and successful project environments.

Chapter 7: AI in Closure and Evaluation

AI revolutionizes lessons learned extraction by processing vast amounts of project data with unprecedented speed and accuracy. Unlike manual reviews that may miss critical patterns, AI algorithms systematically analyze project documentation, communications, and performance metrics to identify key insights. This automated approach significantly reduces the time required for post-project analysis while improving the quality of extracted lessons. AI can process multiple projects simultaneously, identifying cross-cutting themes and best practices that might otherwise remain hidden. The technology excels at uncovering both successful strategies worth replicating and problem areas requiring attention. By leveraging machine learning, these systems continuously improve their analysis capabilities, becoming more sophisticated at identifying subtle patterns and delivering actionable recommendations for future project success.

One of AI's most powerful capabilities in project closure is automatically identifying recurring themes and issues across multiple projects. Natural Language Processing enables AI to analyze text-based project documents, meeting notes, and status reports to extract meaningful patterns. This pattern recognition capability allows organizations to spot systemic issues that may not be apparent when examining individual projects in isolation. For example, AI might identify that resource allocation problems consistently emerge in the third quarter of projects, or that certain communication patterns correlate with project success. Early detection of these recurring themes enables proactive problem-solving, allowing organizations to address root causes rather than symptoms. This systematic approach to issue identification supports more strategic decision-making and helps prevent the repetition of costly mistakes across future initiatives.

Chapter 7: AI in Closure and Evaluation

AI transforms organizational knowledge management by systematically collecting, organizing, and sharing lessons learned across all projects. Traditional knowledge retention often relies on individuals remembering to document and share insights, leading to knowledge silos and lost learning opportunities. AI systems automatically capture lessons learned from diverse sources and organize them in easily searchable, categorized databases. The technology ensures relevant stakeholders receive tailored knowledge recommendations based on their roles, current projects, and historical interests. This efficient knowledge sharing mechanism enhances decision-making quality and accelerates project delivery by leveraging proven successful approaches. Organizations implementing AI-driven knowledge retention report improved project outcomes, reduced repeated mistakes, and enhanced capability development across teams. The result is a continuous learning culture that builds institutional knowledge and competitive advantage over time.

AI-powered KPI and OKR analysis provides continuous, real-time monitoring of project performance metrics, replacing manual tracking with intelligent automation. Traditional performance monitoring often suffers from delayed reporting and inconsistent data collection, making it difficult to identify issues before they become critical. AI tools integrate with project management systems to continuously collect and analyze performance data, providing instant visibility into project health. Real-time analysis enables quick identification of performance gaps and emerging opportunities, allowing project managers to make timely adjustments. This proactive approach helps ensure projects stay aligned with defined goals and organizational objectives. The continuous monitoring capability also supports more accurate forecasting and resource planning, as AI systems learn from historical patterns to predict future performance trends and potential challenges.

Predictive analytics represents one of AI's most valuable contributions to project planning and risk management. By analyzing historical project data, current performance metrics, and external factors, AI creates predictive models that forecast potential risks and likely outcomes. This capability enables organizations to set more realistic and achievable goals based on data-driven insights rather than optimistic assumptions. Predictive models can identify projects likely to experience delays, budget overruns, or quality issues, allowing for early intervention strategies. The technology also supports more effective risk mitigation by highlighting potential problems before they materialize. Organizations using predictive analytics report significantly improved project success rates, better resource utilization, and more accurate timeline estimates. This forward-looking capability transforms project management from reactive problem-solving to proactive strategic planning.

Complex project data becomes accessible and actionable through AI-powered visualization tools that convert overwhelming datasets into clear, intuitive graphics. Traditional reporting often presents stakeholders with dense spreadsheets and lengthy documents that obscure key insights. AI visualization tools automatically generate charts, dashboards, and interactive displays that highlight critical performance trends and patterns. These visual representations enable stakeholders to quickly grasp project status, identify emerging issues, and understand performance trajectories. The technology supports strategic decision-making by presenting data in formats that facilitate rapid comprehension and analysis. Customizable dashboards ensure different stakeholder groups receive relevant visualizations tailored to their needs and responsibilities. This enhanced data accessibility democratizes project insights, enabling more informed decisions at all organizational levels and improving overall project governance effectiveness.

Automated report generation transforms project closure from a time-consuming manual process into an efficient, accurate, and consistent practice. Traditional closure reporting often involves significant manual effort, creating delays and introducing opportunities for errors or omissions. AI systems analyze comprehensive project data to create detailed closure reports automatically, ensuring all critical elements are captured and properly documented. These systems can generate multiple report formats tailored to different audiences, from executive summaries to detailed technical analyses. Automated generation significantly reduces the time required for project closure while improving report quality and consistency. The technology ensures compliance with organizational standards and regulatory requirements by incorporating necessary templates and validation checks. This efficiency gain allows project teams to focus on strategic analysis and planning rather than administrative documentation tasks.

AI-driven file organization revolutionizes project documentation management through intelligent classification and categorization systems. Traditional project file management often results in inconsistent folder structures, misplaced documents, and difficulty locating critical information. AI-powered classification systems automatically analyze document content, metadata, and context to organize files according to predefined taxonomies or learned organizational patterns. This automated organization ensures consistent structure across all projects while making documents easily discoverable through intelligent search capabilities. Team members can quickly access needed information without extensive searching, improving productivity and reducing frustration. The system continuously learns from user interactions and feedback, refining its classification accuracy over time. Enhanced accessibility supports better project management decisions by ensuring relevant historical information is readily available when needed.

AI validation systems ensure project documentation accuracy and consistency by automatically checking documents against established standards and requirements. Manual document review processes are time-consuming and prone to human error, potentially missing critical omissions or inconsistencies. AI tools systematically validate document completeness, format compliance, and content accuracy, flagging potential issues for human review. These systems can cross-reference information across multiple documents to identify discrepancies or missing elements. Consistent validation processes enhance project record reliability, supporting better audit outcomes and regulatory compliance. The technology also identifies opportunities for documentation improvement, suggesting updates to templates and standards based on recurring issues. Organizations implementing AI validation report improved document quality, reduced compliance risks, and enhanced confidence in their project records. This systematic approach to quality assurance strengthens organizational project management capabilities.

Natural Language Processing enables AI systems to analyze qualitative team feedback with unprecedented depth and consistency. Traditional feedback analysis relies on manual interpretation, which can be subjective, time-consuming, and prone to bias. NLP technology processes written feedback, survey responses, and communication data to extract meaningful insights about team sentiment, concerns, and suggestions. The system can categorize feedback themes, identify emotional undertones, and track sentiment changes over time. This automated analysis reveals team dynamics and communication patterns that might not be apparent through manual review. NLP also enables processing of large volumes of feedback quickly, ensuring comprehensive coverage of team input. The technology provides objective analysis that complements human interpretation, offering data-driven insights into team perceptions and experiences throughout project lifecycles.

Sentiment analysis provides project leaders with powerful tools for monitoring team morale and proactively addressing concerns before they impact project outcomes. Traditional morale assessment often relies on periodic surveys or informal observations, potentially missing early warning signs of team dissatisfaction. AI-powered sentiment analysis continuously monitors team communications, feedback, and interaction patterns to detect emotional trends and morale shifts. This real-time monitoring enables leaders to identify emerging issues quickly and implement targeted interventions to maintain positive team dynamics. The technology can track sentiment across different project phases, team subgroups, and communication channels, providing comprehensive visibility into team emotional health. Early detection capabilities support proactive leadership responses, helping maintain productive work environments and prevent minor concerns from escalating into major problems that could derail project success.

Chapter 7: AI in Closure and Evaluation

Integration of sentiment analysis results into continuous improvement processes creates a feedback loop that enhances both team wellbeing and project outcomes. Rather than treating sentiment data as isolated information, organizations can incorporate these insights into their project evaluation and planning processes. Sentiment-informed evaluations provide a more complete picture of project performance by considering both quantitative metrics and qualitative team experiences. This holistic approach ensures that project success is measured not only by deliverable completion but also by team satisfaction and sustainable working practices. Using sentiment insights helps align team wellbeing with project objectives, creating conditions for better long-term performance and reduced turnover. The continuous improvement aspect enables organizations to refine their approaches based on team feedback patterns, building more effective and supportive project environments that consistently deliver superior results.

AI implementation in project closure and evaluation delivers transformative benefits across multiple dimensions of organizational capability. The technology enables deeper project insights through advanced data analysis that reveals patterns and opportunities invisible to manual review processes. Improved documentation accuracy and completeness ensure thorough project records that support better decision-making and compliance. Enhanced team engagement results from more responsive leadership practices informed by sentiment analysis and feedback interpretation. Most significantly, AI adoption supports continuous organizational growth by establishing systematic learning processes that build institutional knowledge and competitive advantage. Organizations embracing these technologies report improved project success rates, enhanced team satisfaction, and stronger capability development. The compound benefits of AI implementation create sustainable improvements that accelerate organizational maturity and market competitiveness in an increasingly complex business environment.

Chapter 8: Predictive Analytics for Projects

Predictive analytics represents a paradigm shift in project management, moving us from reactive to proactive decision-making. At its core, predictive analytics uses both historical project data and current performance indicators to forecast future outcomes with remarkable accuracy. This capability enables project managers to make informed decisions before problems arise, rather than responding to issues after they've already impacted the project. The significance lies in four key areas: enhanced forecasting accuracy, proactive decision-making capabilities, early risk identification and mitigation, and optimized resource allocation. By implementing predictive analytics, organizations can substantially improve their project success rates.

Effective predictive analysis relies on diverse data sources that paint a complete picture of project performance. Project management data forms the foundation, including detailed schedules, budget allocations, resource usage patterns, and milestone completion rates. Risk and external data add crucial context through risk logs, stakeholder feedback, market conditions, and regulatory changes. The distinction between structured and unstructured data is critical for model accuracy. Structured data includes numerical metrics, dates, and categorical information, while unstructured data encompasses emails, meeting notes, change requests, and project communications. Combining both data types significantly enhances the precision and reliability of predictive models.

Implementing predictive analytics delivers substantial benefits while presenting manageable challenges. Improved forecasting accuracy enables better strategic planning and more reliable project timelines. Early risk detection allows teams to address potential problems before they escalate into major issues. Enhanced resource management ensures optimal allocation and utilization across projects. However, implementation challenges must be addressed systematically. Data quality issues require establishing robust data collection and validation processes. System integration challenges need careful planning and potentially significant infrastructure updates. Stakeholder acceptance requires change management, training, and demonstrating clear value. Success depends on addressing these challenges proactively while maintaining focus on the substantial benefits.

Early warning systems represent the first line of defense against project failures, providing continuous monitoring of critical project indicators.

These systems operate by establishing baseline metrics and threshold values, then continuously comparing actual performance against these benchmarks. When deviations occur, the system immediately alerts project managers, enabling rapid response before issues compound. The key advantage is preventing negative impacts through early intervention rather than damage control. Continuous monitoring ensures no critical indicators are overlooked, while automated alerting systems provide timely notifications. This proactive approach significantly reduces the likelihood of project failure and helps maintain stakeholder confidence throughout the project lifecycle.

Successful early warning systems monitor four critical categories of indicators. Schedule slippage appears through delays in milestone completion, extended task durations, and critical path violations. These signals indicate potential timeline risks requiring immediate attention. Cost variances manifest as differences between planned and actual expenditures, budget overruns in specific categories, and unexpected expense increases. Resource conflicts arise from competing demands for limited resources, skill shortages, and allocation inefficiencies. Quality defects include deliverable issues, rework requirements, and stakeholder satisfaction concerns. Monitoring these indicators systematically provides comprehensive project health visibility and enables targeted interventions to maintain project success trajectories.

Organizations implementing effective early warning systems report significant improvements in project outcomes. These systems reduce project delays by an average of 25-40% through timely identification and resolution of potential issues. Cost overruns decrease substantially when warning systems enable proactive budget management and resource reallocation. Predictive analytics enhances decision-making quality by providing data-driven insights rather than intuition-based choices. Successful organizations demonstrate measurable improvements in project delivery efficiency, stakeholder satisfaction, and team productivity. The practical value becomes evident through improved project completion rates, reduced emergency interventions, and more accurate project forecasting. These success stories provide compelling evidence for predictive analytics adoption across industries.

Forecasting project delays requires sophisticated analytical approaches that identify patterns in historical data and current project performance. Statistical analysis forms the foundation, using regression models, time series analysis, and correlation studies to identify delay patterns across similar projects. Machine learning models enhance prediction accuracy by learning from complex data relationships and continuously improving their forecasting capabilities. These algorithms can process vast amounts of project data, identifying subtle patterns that human analysts might miss. Trend analysis provides ongoing monitoring of project trajectories, comparing current performance against planned schedules and historical norms. Together, these techniques create robust delay prediction systems that enable proactive schedule management.

Cost overrun prediction employs multiple data-driven methodologies to assess financial risks before they materialize. Regression models analyze relationships between project characteristics and historical cost performance, identifying factors that consistently lead to budget overruns. These models can predict overrun probability and magnitude based on current project conditions. Bayesian networks model complex probabilistic relationships among cost drivers, providing sophisticated risk assessment capabilities that account for uncertainty and interdependencies. Simulation approaches use Monte Carlo methods and scenario modeling to evaluate thousands of possible project outcomes, estimating likelihood and impact of various cost overrun scenarios. These techniques provide comprehensive cost risk assessment and enable proactive budget management strategies.

Integrating predictions into project planning transforms reactive management into proactive control. Incorporating delay predictions enables dynamic schedule adjustments, resource reallocation, and risk mitigation strategies before problems occur. This integration allows project managers to build contingency plans and adjust timelines based on predicted risks rather than waiting for delays to manifest. Cost prediction integration provides similar benefits for budget management, enabling proactive financial controls and resource optimization. The result is improved project control through dynamic planning that adapts to changing conditions and predicted outcomes. This approach enhances decision-making reliability and significantly improves project success rates by enabling preventive rather than corrective actions.

Resource burn rate represents the speed at which project resources are consumed relative to planned consumption patterns. Understanding burn rate is crucial because it directly impacts both project timeline and budget outcomes. High burn rates can accelerate project completion when resources are abundant but may lead to shortages and delays when resources become constrained. The burn rate affects project progress by determining how quickly work can be completed and deliverables produced. It directly influences project costs through resource expenditure rates and efficiency metrics. Monitoring burn rate provides early indicators of resource management issues, budget risks, and timeline concerns. Effective burn rate management ensures optimal resource utilization while maintaining project quality and timeline objectives.

Analytical models for burn rate forecasting employ multiple sophisticated approaches to predict future resource consumption patterns. Time series forecasting analyzes historical consumption data to identify trends, seasonal patterns, and cyclical variations in resource usage. These models excel at capturing regular patterns and extrapolating them into future periods. Machine learning models enhance forecasting accuracy by incorporating multiple variables and learning from complex relationships between project characteristics and resource consumption. Neural networks and ensemble methods can process vast datasets to identify subtle patterns affecting burn rate. Predicting future consumption enables proactive resource planning, early identification of potential shortages, and optimized allocation strategies that prevent bottlenecks and delays.

Proactive resource allocation strategies leverage burn rate forecasting to optimize project performance and prevent resource-related delays. Burn rate forecasting enables project managers to anticipate resource needs, identify potential shortages, and plan procurement or reallocation activities well in advance. Resource allocation adjustment becomes data-driven rather than reactive, allowing managers to redistribute resources before critical shortages occur. This approach ensures efficient project progress by maintaining optimal resource levels throughout the project lifecycle. Avoiding bottlenecks requires continuous monitoring and adjustment based on predicted consumption patterns. Proactive adjustments help maintain smooth workflow, prevent delays caused by resource constraints, and ensure on-time delivery while maintaining quality standards and stakeholder satisfaction.

AI techniques revolutionize change impact analysis by providing sophisticated methods for understanding how modifications affect project outcomes. Natural Language Processing enables analysis of textual project documents, change requests, and stakeholder communications to identify potential impacts on project scope, schedule, and resources. NLP can extract insights from unstructured data that traditional analysis methods might overlook. Neural networks model complex relationships between changes and their cascading effects throughout project systems. These models can predict how seemingly minor changes might trigger significant downstream impacts. Simulation techniques create virtual environments for testing change scenarios, allowing teams to explore multiple "what-if" situations and their potential consequences before implementing actual changes.

Modeling and simulating change scenarios provides powerful capabilities for understanding potential impacts before implementation. AI-driven simulation creates detailed virtual representations of project environments, enabling teams to test various change scenarios and predict their outcomes with high accuracy. These simulations can model complex interactions between different project elements, stakeholders, and external factors. Outcome evaluation through simulation allows project teams to assess potential risks, benefits, and unintended consequences before committing to changes. This approach significantly reduces the likelihood of negative surprises and enables more informed decision-making. Teams can explore multiple alternatives, compare outcomes, and select optimal approaches based on predicted results rather than guesswork or limited analysis.

Real-world applications of AI-driven change impact analysis demonstrate significant benefits across diverse industries and project types. Change impact analysis helps organizations understand the full ramifications of proposed modifications, reducing unexpected issues and improving change success rates. Construction, software development, manufacturing, and healthcare projects all benefit from these analytical capabilities. Improved adaptability results from enhanced understanding of change impacts, enabling organizations to respond more quickly and effectively to new information, market conditions, and stakeholder requirements. Higher project success rates emerge from better change management, reduced surprises, and more informed decision-making. Organizations implementing AI-driven change analysis report improved stakeholder satisfaction, reduced project risks, and enhanced ability to deliver successful outcomes even in dynamic environments.

In this chapter, we've explored how predictive analytics transforms project management from reactive to proactive practice. Early risk detection capabilities enable identification and mitigation of potential issues before they impact project outcomes, significantly improving success rates. Accurate forecasting through AI-powered tools enhances project planning accuracy, resource allocation efficiency, and timeline reliability. Informed change management leverages AI insights to support better decision-making and more effective change implementation strategies. The integration of these predictive analytics capabilities creates a comprehensive approach to project management that reduces risks, improves outcomes, and enhances stakeholder satisfaction. Organizations adopting these approaches gain competitive advantages through superior project delivery and management capabilities.

Chapter 9: Conversational AI in PM

We begin with the fundamentals of conversational AI in project management. This section establishes the conceptual foundation for understanding how artificial intelligence can revolutionize project communication and coordination. We'll define key terms, explore core technologies, and examine why these tools are becoming essential for modern project management success.

Conversational AI encompasses technologies that understand and respond to human language naturally and effectively. The core technologies include chatbots and virtual assistants that enable automated, natural, and interactive communication between team members and systems. The primary purpose is automating communication tasks while enhancing user experience through natural interactions. These systems process natural language inputs, interpret intent, and provide contextually appropriate responses. Unlike traditional software interfaces that require specific commands or navigation, conversational AI allows users to interact using everyday language, making technology more accessible and intuitive for project teams.

Conversational AI proves highly relevant to project management through two primary benefits. First, it automates routine queries, freeing project managers from repetitive questions about schedules, deadlines, and status updates. This automation allows managers to focus on complex strategic tasks requiring human judgment and creativity. Second, it enhances team collaboration by facilitating timely communication among team members, improving coordination and decision-making processes. When team members can instantly access project information or get quick answers to common questions, the entire project workflow becomes more efficient. This leads to reduced delays, better resource allocation, and improved project outcomes through enhanced communication flow.

Conversational AI applications in project management span four key areas. Virtual project assistants help manage tasks and schedules, boosting team productivity through intelligent automation. Real-time stakeholder support provides immediate assistance, improving communication efficiency and reducing response times. Voice command interfaces enable hands-free control of project management tools, supporting seamless operations even when users are multitasking or have accessibility needs. Multilingual chatbots foster inclusivity by supporting diverse language communication in global projects. Each application addresses specific project management challenges while contributing to overall efficiency improvements. These technologies work together to create comprehensive communication ecosystems that support modern project requirements.

The second major area focuses on chatbots as project assistants. These intelligent systems serve as virtual team members, handling routine tasks and providing consistent support across project lifecycles. We'll examine how they automate essential project functions and integrate seamlessly into existing workflows.

Chatbots excel at automating routine project tasks across three critical areas. Task status updates occur automatically, keeping projects on track without requiring manual input from team members or project managers. This ensures consistent, timely updates and reduces the risk of overlooked status changes. Automated reminders ensure deadlines and meetings are never missed, sending notifications based on project schedules and individual responsibilities. Progress tracking happens in real-time, enabling efficient monitoring and reporting without constant manual data entry. These automated functions free project managers from administrative overhead, allowing them to focus on strategic planning, problem-solving, and team leadership activities that require human insight and decision-making capabilities.

Meeting scheduling and deliverable tracking represent key chatbot capabilities. Calendar coordination allows efficient scheduling across multiple calendars, eliminating conflicts and delays through intelligent availability checking. The chatbot can propose meeting times that work for all participants and automatically send calendar invitations. Meeting arrangement becomes automated, reducing manual effort while ensuring timely organization of project discussions. This includes preparing agendas, notifying participants, and setting up virtual meeting rooms. Deliverable monitoring involves tracking project deadlines to monitor deliverables and maintain project schedules effectively. The system can send escalating reminders, flag at-risk deliverables, and provide real-time status updates to stakeholders, ensuring nothing falls through the cracks.

Documentation and reporting benefit significantly from chatbot assistance. Automated document generation reduces manual errors while saving considerable time in creating project documents, status reports, and compliance materials. The system can pull data from multiple sources, format information consistently, and generate reports based on templates. Organized reporting ensures systematic report organization, providing easy access and clarity for stakeholders at all levels. Reports are categorized, tagged, and stored in searchable formats. Timely distribution enables prompt report sharing with stakeholders, enhancing project communication efficiency. Automated distribution lists ensure the right information reaches the right people at the right time, reducing communication gaps and keeping everyone informed about project progress and changes.

AI for stakeholder Q&A represents our third focus area. This technology transforms how project teams handle information requests and stakeholder communication, providing immediate access to project data and reducing response times for common inquiries.

Real-time information retrieval transforms stakeholder engagement through immediate data access and automated status updates. AI tools provide stakeholders with instant access to current project information anytime, eliminating the need to wait for scheduled updates or email responses. This 24/7 availability improves stakeholder satisfaction and keeps everyone informed. Stakeholders can query project status and metrics without waiting for manual reports from project managers or team members. The system can provide real-time dashboards, progress percentages, budget utilization, timeline updates, and resource allocation information. This self-service approach reduces the administrative burden on project teams while ensuring stakeholders have the information they need to make informed decisions and provide timely feedback.

Transparency and communication improvements represent core benefits of conversational AI implementation. The technology enhances transparency by enabling clear and consistent communication across all platforms and touchpoints. Information remains consistent regardless of who asks or when they ask, reducing confusion and miscommunication. AI-powered channels keep all parties informed and engaged, fostering openness and trust among project stakeholders. This continuous communication loop helps identify issues early, facilitates faster decision-making, and builds confidence in project management capabilities. Stakeholders feel more connected to project progress and more confident in outcomes when they have easy access to accurate, up-to-date information whenever they need it.

FAQ response automation delivers significant efficiency gains through three key mechanisms. Automated FAQ responses enable faster information delivery without manual intervention, reducing response times from hours or days to seconds. This immediate availability improves user experience and reduces frustration with communication delays. Reduced delays in providing answers to common questions eliminates bottlenecks in project communication workflows. Team members can get answers quickly and continue their work without waiting for responses. Improved stakeholder satisfaction results from faster, accurate responses that enhance the overall communication experience. When stakeholders can reliably get quick answers to their questions, they develop greater confidence in the project team and are more likely to provide timely feedback and support for project initiatives.

Voice-to-plan interfaces represent an innovative approach to project management interaction. These systems enable hands-free, voice-driven project updates and control, supporting accessibility needs and multitasking scenarios common in modern work environments.

Voice commands transform project plan management through three key capabilities. Quick schedule modification allows rapid changes to project schedules without manual input, saving significant time during project adjustments. Project managers can verbally instruct systems to move deadlines, reschedule meetings, or adjust resource allocations while reviewing documents or participating in discussions. Task assignment by voice enables managers to assign tasks verbally, streamlining communication and reducing delays in task distribution. This eliminates the need to interrupt work flow to access assignment systems. Progress logging efficiency improves through voice-enabled progress updates, enhancing productivity and record accuracy. Team members can update their progress while working, ensuring real-time project visibility without disrupting their focus on deliverable completion.

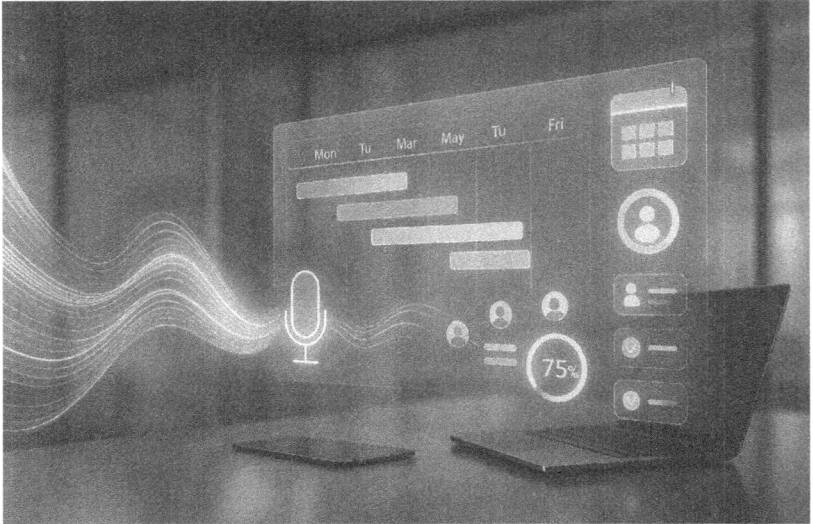

Integration with project management tools enables seamless voice-to-plan functionality. Voice-to-plan systems connect directly with project management software, enabling efficient updates without switching between applications or interfaces. This integration ensures data consistency across all project management tools and maintains audit trails for changes. Real-time document updates occur automatically as voice inputs update project documentation, ensuring accuracy and timeliness. Changes made through voice commands immediately reflect in project plans, schedules, and reporting systems. This integration eliminates the delay between verbal decisions and documented changes, reducing the risk of miscommunication or forgotten updates. The seamless connection between voice input and project systems creates a more efficient workflow for project managers and team members.

Accessibility and hands-free operation provide significant benefits for diverse project teams. Voice interfaces improve accessibility for users with disabilities, making project management more inclusive and compliant with accessibility standards. This ensures all team members can effectively participate in project activities regardless of physical limitations. Hands-free voice control allows multitasking and seamless management of project workflows, particularly valuable for field workers, mobile professionals, or anyone whose hands are occupied with other tasks. Project managers can update schedules while reviewing physical deliverables, assign tasks during site visits, or log progress while driving between meetings. This flexibility improves productivity and ensures project information stays current even in challenging work environments.

Multilingual PM chat assistants address the growing need for inclusive communication in global project teams. These systems bridge language barriers and cultural differences that can impact project success in diverse, international environments.

Language barrier management represents a critical capability for global project success. Multilingual chatbots support multiple languages, enabling smooth communication in diverse global teams where members may have different native languages. These systems can switch between languages dynamically or maintain parallel conversations in different languages simultaneously. Using multilingual chatbots minimizes language barriers and prevents communication errors that could lead to project delays or quality issues. The technology ensures that language differences don't become obstacles to effective collaboration. Clear communication in preferred languages helps team members express complex ideas, concerns, and suggestions more effectively, leading to better problem-solving and innovation within project teams.

Real-time translation and localization capabilities enhance global project coordination. AI assistants deliver immediate translations, facilitating seamless communication across languages without requiring external translation services or delays. This real-time capability maintains conversation flow and project momentum. Cultural adaptation features adjust communication styles to suit diverse local cultures and customs, ensuring messages are not only translated but culturally appropriate. This includes adjusting formality levels, communication patterns, and contextual references. Enhanced team cohesion results from improved clarity and understanding, as AI aids in strengthening teamwork and collaboration. When team members can communicate effectively regardless of language differences, trust builds more quickly and project teams function more cohesively.

Inclusivity and collaboration improvements deliver measurable project benefits. Accessible communication through multilingual chatbots enables participation from all team members regardless of language differences, ensuring diverse perspectives contribute to project success. This inclusivity often leads to more creative solutions and better risk identification. Fostering inclusivity creates welcoming environments where diverse voices are heard and valued, improving team morale and retention. When team members feel comfortable communicating in their preferred language, they're more likely to share important insights and concerns. Improved project outcomes result from inclusive communication that enhances team collaboration, leading to better decision-making and more successful project results through leveraging the full potential of diverse global teams.

In conclusion, conversational AI transforms project management through three key impact areas. Efficiency enhancement occurs as AI automates routine tasks, significantly improving project management efficiency and saving valuable time that can be redirected to strategic activities. This automation reduces administrative overhead and accelerates project workflows. Improved collaboration results from AI-driven communication tools that foster better teamwork and coordination among project members, regardless of location, time zone, or language differences. These tools create more connected, responsive project environments. Inclusivity through AI supports diverse project environments by accommodating different communication needs, accessibility requirements, and cultural preferences. This inclusive approach leads to stronger teams, better outcomes, and more sustainable project management practices in our increasingly global business environment.

Chapter 10: Integrating AI into PMO Practice

PMO dashboards serve as the central nervous system of project management operations, providing critical visibility into project status, resource allocation, and performance metrics. Traditional dashboards often require manual data entry and updates, creating bottlenecks and potential for human error. AI enhancement transforms these static reporting tools into dynamic, intelligent systems that can automatically aggregate data from multiple sources, provide real-time insights, and generate predictive analytics. This transformation enables PMO teams to shift from data collection to strategic decision-making, ultimately improving project outcomes and organizational efficiency.

AI revolutionizes dashboard functionality through three key capabilities that address traditional PMO challenges. Automated data aggregation eliminates the time-consuming manual process of collecting information from disparate project management tools, reducing errors and ensuring data consistency. Real-time information updates provide decision-makers with current project status, enabling rapid response to emerging issues. Enhanced decision support leverages AI algorithms to analyze aggregated data and surface actionable insights, helping PMO teams identify trends, predict potential problems, and make informed strategic decisions. These capabilities collectively transform dashboards from passive reporting tools into active management platforms.

Predictive analytics represents one of AI's most powerful contributions to project tracking, offering PMO teams unprecedented foresight into project trajectories. Risk anticipation uses machine learning algorithms to analyze historical project data, team performance patterns, and external factors to identify potential issues before they materialize. Delay prediction models forecast timeline slippages based on current project velocity and resource constraints, enabling proactive schedule adjustments. Bottleneck identification analyzes workflow patterns to pinpoint process inefficiencies and resource constraints that could impede project progress. These predictive capabilities transform reactive project management into proactive strategic planning, significantly improving project success rates.

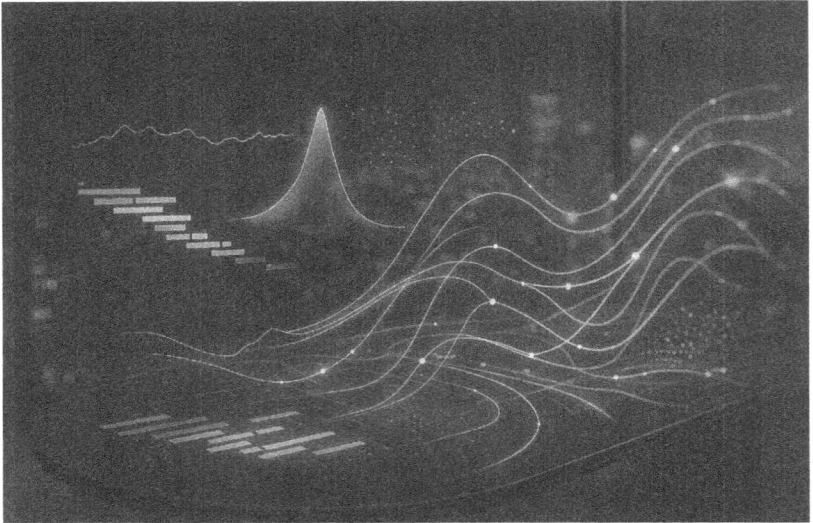

AI-powered visualization transforms complex project data into intuitive, actionable insights that stakeholders can quickly understand and act upon. Traditional KPI reporting often overwhelms decision-makers with static charts and tables that require significant interpretation. AI-driven visualization tools automatically generate dynamic, interactive displays that highlight critical information and trends. Customizable visual insights adapt to specific stakeholder needs, ensuring executives see strategic overviews while project managers access detailed operational metrics. Efficient project monitoring becomes possible through real-time visual updates that immediately reflect project status changes, enabling faster decision-making and more effective stakeholder communication throughout the project lifecycle.

Portfolio management represents the strategic heart of PMO operations, where decisions about resource allocation, project prioritization, and risk management directly impact organizational success. Traditional portfolio management often relies on subjective assessments and manual processes that can miss critical patterns or opportunities. AI-driven portfolio management transforms this landscape by providing objective, data-driven insights that optimize resource utilization and maximize portfolio value. Through intelligent algorithms and predictive modeling, AI enables PMOs to make more informed decisions about which projects to pursue, how to allocate resources, and when to pivot strategies based on changing business conditions.

AI transforms project prioritization from subjective decision-making to objective, data-driven analysis that maximizes organizational value. Intelligent project prioritization algorithms evaluate multiple factors simultaneously, including strategic alignment, resource requirements, risk profiles, and potential return on investment. These systems can process vast amounts of historical project data to identify patterns that predict success likelihood. Optimal resource allocation uses AI to match project requirements with available capabilities, ensuring the right people work on the right projects at the right time. This approach maximizes portfolio value by ensuring resources are directed toward projects with the highest probability of success and strategic impact.

AI-powered risk assessment provides continuous, comprehensive monitoring that surpasses traditional periodic risk reviews. Continuous risk monitoring uses machine learning algorithms to analyze project data streams in real-time, detecting patterns and anomalies that might indicate emerging risks. These systems can identify subtle correlations between seemingly unrelated factors that human analysis might miss. Timely risk mitigation becomes possible through early warning systems that alert PMO teams to developing issues before they impact project outcomes. This proactive approach to risk management enables organizations to address problems while they're still manageable, significantly reducing the likelihood of project failures and cost overruns.

Machine learning models bring scientific rigor to project outcome prediction, transforming intuitive project management into evidence-based decision-making. These models analyze historical project data, team performance metrics, and environmental factors to predict success probabilities with remarkable accuracy. Supporting decision-making becomes more robust when PMO teams have access to quantitative predictions about project trajectories, enabling them to make informed adjustments to scope, resources, or timelines. Improving delivery outcomes happens naturally when teams can anticipate challenges and opportunities, allowing for proactive strategy adjustments that optimize project results. This predictive capability represents a fundamental shift from reactive to proactive project management.

Governance and auditing form the backbone of effective project management, ensuring compliance with organizational standards, regulatory requirements, and industry best practices. Traditional governance processes often rely on manual reviews, periodic audits, and subjective assessments that can miss critical issues or create bottlenecks. AI integration into governance and auditing processes automates routine compliance checks, provides continuous monitoring capabilities, and delivers objective performance assessments. This transformation not only improves the accuracy and consistency of governance activities but also frees up human resources to focus on strategic oversight and relationship management rather than administrative tasks.

AI revolutionizes compliance management by automating routine verification processes and generating comprehensive reports with minimal human intervention. Automated compliance verification systems continuously monitor project activities against established standards, immediately flagging deviations or potential issues. This real-time monitoring eliminates the lag time inherent in traditional periodic audits, enabling immediate corrective action when problems arise. Automated reporting generates precise, standardized compliance documentation that reduces manual workload and improves audit readiness. These systems maintain consistent documentation standards across all projects, ensuring that compliance information is accurate, complete, and readily available for internal reviews or external audits, significantly reducing compliance-related risks.

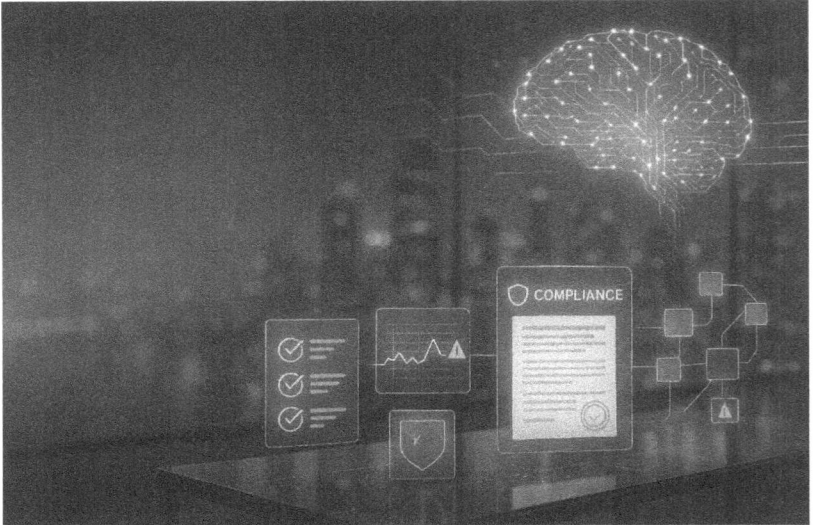

AI-powered monitoring systems provide unprecedented visibility into governance policy adherence across project portfolios, transforming compliance from a reactive to a proactive discipline. Continuous monitoring capabilities track project activities against established governance frameworks in real-time, ensuring that policy violations are detected immediately rather than during periodic reviews. Early deviation detection algorithms identify potential compliance issues before they become serious problems, enabling swift intervention and course correction. Facilitating corrective actions becomes more efficient when AI systems can automatically generate recommended remediation steps based on the type and severity of policy deviations, helping project teams maintain compliance while minimizing disruption to project progress.

AI auditing tools transform organizational transparency and accountability by providing objective, data-driven insights into project performance and governance compliance. AI-driven transparency emerges through comprehensive analysis of project data that reveals patterns, trends, and correlations that might not be apparent through traditional auditing methods. Fostering stakeholder trust becomes easier when audit results are based on objective algorithmic analysis rather than subjective human judgment. Supporting accountability mechanisms through AI enables organizations to maintain consistent standards across all projects while providing clear, evidence-based performance evaluations. These tools create an environment where accountability is built into the project management process rather than imposed through external oversight.

Scaling AI implementation across multiple projects represents both the greatest opportunity and the most significant challenge in AI-driven project management transformation. While individual project successes demonstrate AI's potential, realizing organization-wide benefits requires systematic approaches to technology adoption, change management, and cultural transformation. Successful scaling involves standardizing AI tools and processes, ensuring interoperability across systems, and developing organizational capabilities to support AI-enhanced project management. This transformation extends beyond technology implementation to encompass training, process redesign, and cultural adaptation that enables organizations to fully leverage AI's transformative potential across their entire project portfolio.

Successful AI adoption across multiple projects requires systematic approaches that ensure consistency, efficiency, and organizational alignment. Establishing clear goals provides the foundation for AI implementation by defining specific objectives and success metrics that guide technology selection and deployment strategies. Standardizing AI tools across the organization improves efficiency and reduces complexity by creating common platforms and processes that teams can easily learn and use. Ensuring interoperability between AI systems and existing project management tools facilitates smooth integration and data flow, preventing the creation of isolated technology silos that can reduce overall effectiveness and increase maintenance costs.

Change management and comprehensive training form the cornerstone of successful AI integration, addressing the human factors that often determine technology adoption success or failure. Team preparation through training builds the technical skills and confidence necessary for effective AI tool utilization while addressing concerns about job security and role changes. Addressing resistance requires proactive communication strategies that help teams understand AI's benefits while providing support during the transition period.

Chapter 10: Integrating AI into PMO Practice

Fostering an AI culture involves creating an environment that encourages experimentation, learning from failures, and continuous improvement. This cultural transformation enables organizations to maximize AI's benefits while maintaining team morale and engagement throughout the implementation process.

Measuring and scaling AI impact requires systematic approaches to performance evaluation and continuous improvement that demonstrate value and guide future investments. Continuous impact measurement involves establishing baseline metrics before AI implementation and regularly tracking improvements in project success rates, efficiency gains, and cost reductions. Process efficiency gains should be quantified through specific metrics such as time savings, error reduction, and resource optimization to demonstrate tangible benefits. Scaling AI practices becomes more effective when organizations use data-driven insights to identify successful implementations that can be replicated across other projects and departments, ensuring that lessons learned are captured and applied systematically.

AI integration into PMO practices delivers transformative benefits across four critical dimensions that collectively revolutionize project management effectiveness. Enhanced data management capabilities enable organizations to harness the full value of their project information through automated collection, processing, and analysis. Improved decision-making results from AI's ability to provide objective, data-driven insights that support more accurate and timely strategic choices. Enhanced governance and scalability emerge through automated compliance monitoring and standardized processes that can grow with organizational needs. Boosting project success and agility becomes achievable when AI tools provide predictive insights and automated processes that enable faster response to changing conditions and emerging opportunities.

Chapter 11: Building an AI-Ready PM Culture

An AI-ready mindset in project management requires three
fundamental pillars. First, embracing innovation means fostering
openness to cutting-edge technologies and committing to continuous
improvement in all project management activities. Second, data-driven
decision-making becomes essential as project managers must leverage
advanced analytics and AI-generated insights to make more informed,
effective choices. Third, flexibility and adaptability are crucial
characteristics that enable teams to integrate AI tools smoothly while
adapting existing workflows to accommodate new technological
capabilities and changing project requirements.

Leadership support forms the cornerstone of successful AI initiatives
within any organization. Leadership endorsement serves as a powerful
signal of organizational commitment, driving AI initiatives forward and
encouraging widespread adoption across teams. Effective resource
allocation becomes possible when strong leadership ensures that
adequate budget, personnel, and technological resources are dedicated
to AI project success. Additionally, visible leadership support builds
organizational confidence, empowering teams to embrace AI
integration with enthusiasm and reducing resistance to technological
change throughout the transformation process.

Fostering collaboration and continuous learning creates the foundation
for sustained AI success. Teamwork encouragement involves
promoting active collaboration among team members, which
strengthens AI project outcomes and facilitates knowledge sharing

across departments. Ongoing education ensures that project teams remain current with the latest AI advancements, techniques, and best practices in the rapidly evolving field. Sharing best practices becomes essential for helping teams implement AI solutions effectively and efficiently across multiple projects, creating organizational learning that benefits the entire company.

Project managers require three essential AI skills for effective leadership in the digital age. Data literacy stands as the foundation, enabling project managers to interpret, analyze, and utilize data effectively for AI-driven project decision-making. AI tool usage represents practical competency in understanding and operating various AI tools and platforms essential for managing AI-enabled project tasks efficiently and successfully. Understanding AI's impact on project delivery ensures that project managers can confidently lead AI projects by grasping how artificial intelligence influences project outcomes, timelines, and resource allocation.

Designing effective training programs requires two key components for maximum impact. Tailored training content should focus specifically on AI competencies essential for effective project management, avoiding generic approaches in favor of targeted skill development that directly applies to your organizational needs. Theory and practice integration ensures that effective training combines comprehensive theoretical knowledge with hands-on practical application, enhancing learning outcomes through real-world experience. This balanced approach helps participants understand both the conceptual foundations and practical implementation of AI in project management contexts.

Encouraging practical application accelerates learning and builds confidence in AI adoption. Hands-on AI experimentation involves encouraging teams to actively use AI tools in controlled environments, supporting effective learning through direct practical experience rather than passive observation. Real-world project application takes this further by applying AI technologies in actual projects, which builds team confidence and eases adoption across the organization. This practical approach helps teams understand the tangible benefits of AI integration while developing the expertise necessary for successful long-term implementation.

Integrating AI tools into existing workflows requires careful planning to maximize benefits while minimizing disruption. Seamless integration ensures that AI tools are incorporated smoothly into current workflows, avoiding operational disruption while maintaining productivity levels during the transition period. Maximized automation benefits result from proper integration, enhancing automation capabilities that improve efficiency and reduce time-consuming manual tasks. Enhanced analytics capabilities emerge through AI integration, enabling advanced data analytics that support better decision-making processes and provide deeper insights into project performance and outcomes.

Optimizing decision-making through data-driven insights transforms how projects are managed and executed. AI-generated insights produce actionable intelligence that empowers faster, more informed decision-making processes throughout the project lifecycle. Improved project outcomes result from using these data-driven insights, which enhance project success rates by enabling better planning, execution, and resource allocation. Enhanced risk management capabilities emerge as data-driven decisions help identify potential risks earlier and develop more effective mitigation strategies, ultimately leading to more successful project completions.

Enhancing efficiency through automation transforms routine project management tasks and strategic focus. AI automates repetitive tasks by handling routine administrative work, data entry, and standard reporting functions, significantly reducing manual effort while increasing overall team productivity. This automation creates opportunities for strategic focus, allowing project managers to dedicate more time and energy to high-value activities like strategy development, complex problem-solving, stakeholder relationship management, and innovation initiatives. The result is a more strategic approach to project management that leverages human creativity and AI efficiency.

Analyzing project challenges systematically reveals optimal opportunities for AI implementation across your organization. Identifying project pain points involves evaluating current project challenges, bottlenecks, and inefficiencies to understand precisely where AI can improve operational efficiency and project outcomes most effectively. AI in forecasting represents a high-impact application area, using artificial intelligence to enhance forecasting accuracy and provide better insights for strategic decision-making. AI for scheduling and resource allocation offers another valuable opportunity, applying AI technologies to optimize project scheduling and ensure efficient resource allocation for improved project success rates.

Selecting high-impact areas for AI implementation ensures maximum return on investment and organizational buy-in. Focus on measurable benefits involves prioritizing areas where AI can provide clear, quantifiable improvements that deliver faster impact and demonstrate tangible value to stakeholders and leadership teams. Tangible AI improvements should be the primary target, implementing AI solutions that deliver visible and immediate enhancements in operations and outcomes. This approach builds momentum for broader AI adoption by creating success stories that validate the investment and encourage further implementation across the organization.

Measuring and communicating early successes builds momentum and validates AI investments across your organization. Tracking AI project metrics involves monitoring key performance indicators that help evaluate the progress, impact, and return on investment of AI projects effectively over time. Sharing results with stakeholders becomes crucial for building trust and encouraging broader acceptance of AI initiatives throughout the organization. Regular communication of successes, lessons learned, and measurable improvements helps maintain support for AI projects while identifying opportunities for expansion and refinement of AI implementation strategies.

Engaging stakeholders in the AI transition ensures organizational alignment and successful implementation. Early stakeholder involvement promotes alignment by engaging key stakeholders from the beginning of the AI journey, which helps address potential concerns proactively and builds consensus around AI objectives. Fostering shared ownership of AI initiatives increases commitment and support for successful implementation by ensuring that stakeholders feel invested in the outcomes. This collaborative approach reduces resistance, improves buy-in, and creates a network of AI champions throughout the organization who can support and advocate for continued AI adoption.

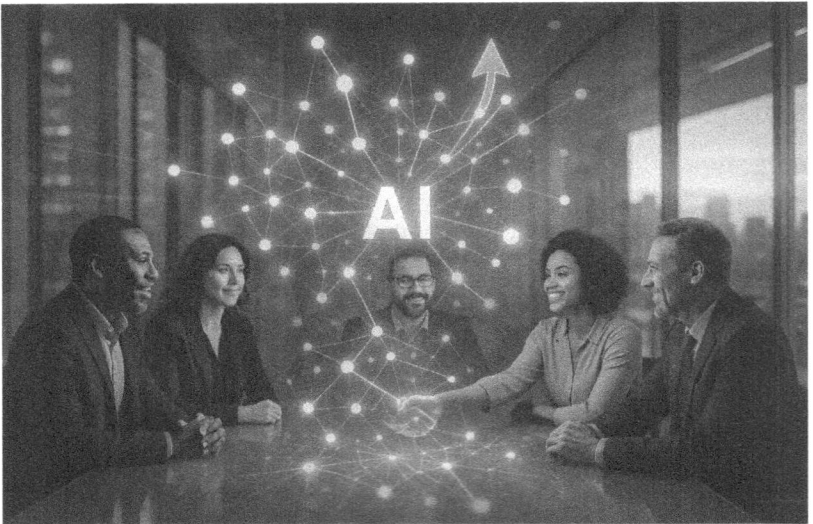

Addressing resistance and building trust requires a multi-faceted communication and education approach. Transparent communication involves maintaining open and clear communication about AI initiatives, which reduces skepticism and promotes better understanding of AI technologies and their benefits. Education on AI provides comprehensive information about AI capabilities, limitations, and applications, building knowledge while easing fears about new technologies. Building trust emerges from combining effective communication and education strategies, which together create public trust in AI innovations and demonstrate the organization's commitment to responsible AI implementation.

Sustaining momentum for long-term transformation requires ongoing commitment and celebration of progress. Continuous support involves providing ongoing assistance, resources, and guidance to help maintain team motivation throughout the AI transformation process, ensuring that momentum doesn't decline over time. Regular updates through frequent progress reports keep everyone informed and engaged in the AI journey, maintaining transparency and accountability. Celebrating milestones by recognizing achievements and successes boosts team morale and encourages continued commitment to long-term AI transformation goals, creating positive reinforcement for sustained effort.

In conclusion, building an AI-enabled project management culture requires attention to four critical areas that work together synergistically. An AI-ready culture that embraces innovation and technological advancement forms the foundation for successful adoption and ongoing innovation in project management. Skills development through strategic investment in employee capabilities ensures effective use and management of AI technologies across all project activities. Process redesign that integrates AI into existing workflows enhances efficiency and improves project outcomes. Finally, effective change management is essential to fully unlock AI's transformative potential in project management, ensuring that your organization maximizes the benefits of AI adoption.

Chapter 12: AI Risks and Governance

Transparency forms the bedrock of trustworthy AI systems. When users understand how AI makes decisions, they develop confidence in the system's recommendations and outputs. This understanding enables meaningful accountability - stakeholders can evaluate whether AI decisions align with organizational values and goals. Conversely, opaque AI systems create significant risks. Without visibility into decision-making processes, organizations cannot identify when systems make errors, exhibit bias, or operate outside intended parameters. This lack of transparency can lead to misplaced trust, inappropriate reliance on AI outputs, and ultimately, poor business outcomes that could have been prevented with better visibility.

Several techniques can improve AI explainability, each serving different stakeholder needs. Model interpretability tools provide technical insights into how algorithms process data and reach conclusions, helping developers and data scientists understand system behavior. Visualizations translate complex mathematical processes into accessible graphics, enabling non-technical stakeholders to grasp AI decision patterns. Simplified surrogate models create approximations of complex systems that are easier to interpret while maintaining reasonable accuracy. The key is matching explainability techniques to audience needs - executives require different explanations than data scientists, and end users need different insights than compliance officers.

Achieving explainable AI presents several significant challenges. Complex AI models, particularly deep neural networks, operate through intricate mathematical processes that are inherently difficult to interpret. There's often a fundamental trade-off between model accuracy and explainability - simpler, more interpretable models may sacrifice performance. Proprietary algorithms add another layer of complexity, as vendors may be reluctant to reveal implementation details. Perhaps most challenging is meeting diverse stakeholder needs - different audiences require different types and levels of explanation. Technical teams need algorithmic details, while business users need outcome interpretations, and regulators may require audit trails demonstrating compliance with specific requirements.

Bias in AI-driven project management tools stems from several common sources that can significantly impact decision quality. Unrepresentative training data creates skewed outputs when the data doesn't reflect the full scope of scenarios the AI will encounter in practice. Historical prejudices embedded in legacy data perpetuate past discriminatory practices, affecting current project decisions. Flawed assumptions built into model design can systematically favor certain project types, teams, or approaches over others. These biases can manifest in resource allocation, risk assessment, timeline estimation, and team assignments, potentially creating unfair advantages or disadvantages that undermine project success and organizational equity.

Poor data quality creates cascading problems throughout AI-driven decision-making processes. Faulty AI predictions based on inaccurate or outdated information lead to misguided project decisions, resource misallocation, and missed opportunities. When stakeholders experience repeated prediction failures, confidence in AI models erodes, creating resistance to data-driven decision making. This reduced confidence can be more damaging than the original data quality issues, as it undermines the entire digital transformation effort. Most critically, poor data quality leads to costly management errors - wrong strategic directions, failed projects, and wasted resources that impact overall business performance and competitive positioning.

Mitigating bias and improving data integrity requires systematic, ongoing efforts across multiple dimensions. Rigorous data auditing involves regular examination of data sources, collection methods, and historical patterns to identify and eliminate hidden biases. This includes statistical analysis to detect skewed distributions and systematic gaps. Diverse data sourcing ensures training datasets represent the full range of scenarios and populations the AI will encounter. Human oversight and feedback create continuous improvement loops where domain experts can identify model outputs that don't align with business reality or ethical standards, enabling iterative refinement of both data and algorithms.

AI failures and hallucinations take various forms, each requiring different management approaches. Misclassifications occur when AI incorrectly categorizes inputs, while erroneous predictions provide inaccurate forecasts or recommendations. Hallucinations represent perhaps the most concerning failure type - AI generating plausible-sounding but completely false information. Data noise disrupts AI learning processes, causing models to identify spurious patterns or miss genuine signals. Model limitations restrict AI's ability to handle edge cases, ambiguous inputs, or scenarios outside training parameters. Adversarial attacks deliberately exploit AI vulnerabilities, while naturally ambiguous inputs can confuse even well-designed systems, leading to unreliable outputs.

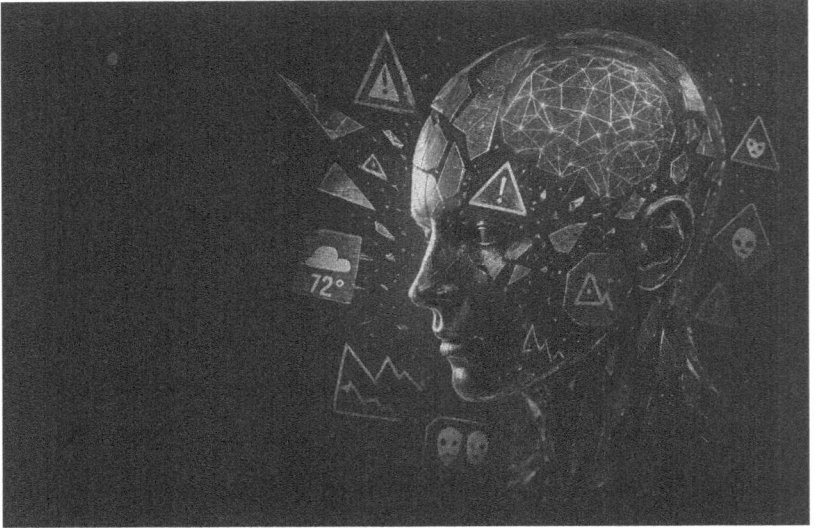

Effective detection and monitoring systems provide early warning of AI failures before they impact business operations. Anomaly detection techniques automatically identify unusual outputs that deviate from expected patterns, flagging potential errors for human review. Confidence scoring helps prioritize attention by indicating how certain the AI is about its outputs - low confidence scores trigger additional verification. Human-in-the-loop verification incorporates expert oversight at critical decision points, combining AI efficiency with human judgment. Continuous performance monitoring tracks AI accuracy, response times, and other key metrics over time, enabling teams to identify degrading performance before it affects end users or business processes.

Robust response and mitigation plans ensure AI failures don't disrupt critical business operations. Fallback procedures provide alternative methods for continuing work when AI systems fail, whether through manual processes, simpler algorithms, or backup systems. Comprehensive error logging captures detailed information about failures for subsequent analysis and prevention. User alerts and notifications ensure stakeholders are promptly informed when AI systems encounter problems, enabling them to adjust their decision-making accordingly. Model retraining with corrected data addresses root causes of failures by improving the underlying algorithms and datasets, creating a continuous improvement cycle that enhances system reliability over time.

Privacy regulations create a complex compliance landscape that AI applications must navigate carefully. The General Data Protection Regulation (GDPR) governs data privacy across the European Union, establishing strict requirements for data processing, user consent, and individual rights. The California Consumer Privacy Act (CCPA) provides similar protections for California residents, emphasizing transparency and consumer control over personal data. Sector-specific standards add additional requirements - healthcare organizations must comply with HIPAA, financial services face various regulatory requirements, and other industries have their own privacy frameworks. Compliance isn't just about avoiding penalties; it builds user trust and demonstrates organizational commitment to ethical AI practices.

Protecting data throughout AI workflows requires multiple layers of security and privacy controls. Data anonymization removes personally identifiable information while preserving the analytical value needed for AI training and operation. Encryption secures data both in transit and at rest, preventing unauthorized access even if systems are compromised. Access controls ensure only authorized personnel can view sensitive information, with role-based permissions limiting data exposure to those who need it for their specific functions. Secure data storage combines physical security, logical controls, and backup procedures to protect information throughout its lifecycle, from collection through processing to eventual deletion.

Maintaining compliance requires ongoing effort and systematic attention to privacy requirements. Regular audits assess current practices against regulatory requirements and identify areas for improvement before they become compliance violations. Privacy impact assessments evaluate new projects and system changes for potential privacy risks, enabling proactive mitigation rather than reactive fixes. Comprehensive employee training ensures all staff understand their privacy responsibilities and know how to handle personal data appropriately. Legal expertise and privacy-by-design principles embed compliance considerations into system architecture and business processes from the beginning, making privacy protection a natural part of operations rather than an afterthought.

Successful AI governance requires balancing multiple critical elements simultaneously. Transparency in AI systems builds user trust and enables effective oversight, but must be balanced with competitive considerations and system complexity. Bias mitigation promotes fairness and equitable outcomes, requiring ongoing attention to data quality, algorithm design, and output monitoring. Failure management ensures AI systems handle errors gracefully and continue providing value even when individual predictions are incorrect. Privacy adherence protects user data and maintains ethical deployment standards while enabling the data processing necessary for AI functionality. Organizations that excel in all four areas create sustainable competitive advantages through trustworthy, reliable AI systems.

Chapter 13: Future Trends and Evolving Standards

The emergence of artificial intelligence in project management represents a fundamental shift in how we approach project planning, execution, and control. AI optimizes project planning by analyzing vast datasets and predicting potential outcomes with unprecedented accuracy, enabling more informed decision-making from the outset. Risk assessment is significantly enhanced through AI's ability to identify patterns and evaluate potential project risks with greater precision than traditional methods. Decision-making automation increases efficiency while reducing human error in routine processes. Perhaps most importantly, predictive analytics powered by AI provides invaluable insights that enhance current project outcomes and inform future planning strategies, creating a continuous improvement cycle that elevates project success rates.

The landscape of project management standards and best practices is undergoing rapid evolution to accommodate AI technologies. Traditional project management standards are being updated to include emerging AI technologies, ensuring that established frameworks remain relevant and effective in modern environments. This integration requires comprehensive updates to best practices, ensuring that AI adoption is both effective and responsible within project contexts. New guidelines for responsible AI use are being developed to maintain ethical standards and efficient implementation within existing project management frameworks. These evolving standards address critical considerations including data privacy, algorithmic bias, transparency requirements, and accountability measures that ensure AI serves as a force multiplier rather than a risk factor in project success.

AI integration spans across all major project management methodologies, enhancing each approach's core strengths while addressing traditional limitations. In traditional methodologies, AI enhances structured planning processes and improves resource allocation through data-driven optimization and predictive modeling. For agile methodologies, AI supports adaptive planning and enables real-time progress monitoring, providing teams with greater flexibility and responsiveness to changing requirements. In hybrid methodologies, AI serves as a bridge between traditional and agile approaches, promoting both responsiveness and efficient project execution. This universal applicability demonstrates AI's versatility and its potential to enhance project outcomes regardless of the chosen methodology, making it a valuable addition to any project manager's toolkit.

PMI has developed comprehensive guidelines and frameworks specifically designed to assist project managers in effectively integrating AI tools into their projects. These structured guidelines provide practical, actionable guidance for identifying appropriate AI applications, selecting suitable tools, and implementing them successfully within existing project workflows. Ethical considerations form a cornerstone of PMI's approach, emphasizing the importance of maintaining strong ethical standards throughout AI adoption processes. Additionally, PMI frameworks incorporate robust governance structures that ensure AI implementation remains controlled, compliant with industry standards, and aligned with organizational objectives. These frameworks address key areas including risk management, stakeholder communication, change management, and performance measurement in AI-enhanced project environments.

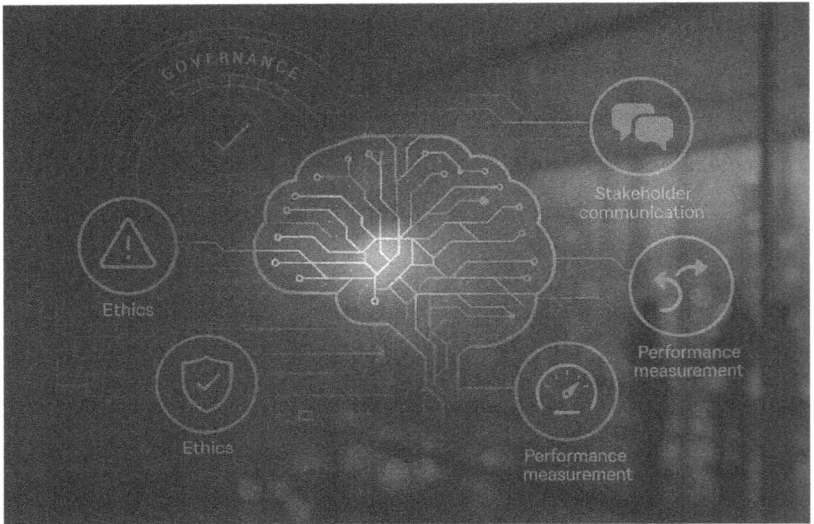

ISO has established comprehensive standards for AI implementation in project management, focusing on ensuring responsible and ethical AI use across diverse industries and organizational contexts. AI governance standards provide clear guidelines for establishing appropriate oversight mechanisms, decision-making processes, and accountability structures for AI systems. Risk management standards specifically address the unique challenges of identifying, assessing, and mitigating risks associated with AI deployment in project management contexts. Interoperability standards promote seamless integration of AI tools within existing project management systems, ensuring that new technologies enhance rather than disrupt established workflows.

Chapter 13: Future Trends and Evolving Standards

These ISO standards provide a foundation for consistent, reliable AI implementation that supports project success while maintaining organizational integrity and compliance requirements.

A comparative analysis reveals that PMI and ISO methodologies offer complementary approaches to AI integration in project management. PMI's methodology emphasizes practical application, focusing on hands-on project management practices that are immediately applicable in real-world situations and facilitate effective AI integration. ISO's framework prioritizes standardization and compliance, ensuring that processes maintain quality and consistency across different projects and organizations. The most effective approach combines both frameworks, creating a comprehensive strategy that enhances AI-driven project management effectiveness and reliability. This integrated approach leverages PMI's practical focus with ISO's standardization emphasis, providing project managers with both the tools needed for immediate implementation and the structural framework necessary for long-term success and organizational scalability.

AI significantly enhances agile planning and execution through three key capabilities that directly support agile principles and practices. Real-time data analysis enables agile teams to analyze information instantly, dramatically improving decision-making speed and quality throughout project iterations and sprint cycles. Predictive insights powered by AI help teams forecast potential challenges and opportunities, enhancing agile responsiveness and enabling proactive rather than reactive management approaches. Support for iterative planning allows AI tools to assist agile teams in adapting plans quickly and effectively based on continuous data feedback and insights from previous iterations. These capabilities align perfectly with agile values of responding to change, customer collaboration, and delivering working solutions, making AI a natural complement to agile methodologies.

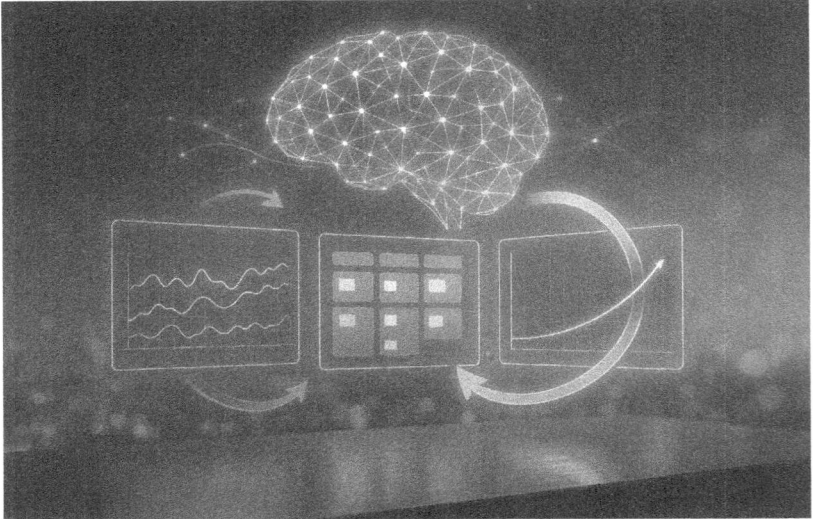

AI augmentation in hybrid project management models addresses the inherent challenge of balancing structure and flexibility that defines hybrid approaches. AI helps achieve optimal balance between rigid traditional processes and flexible agile practices, enabling teams to leverage the best aspects of both methodologies without compromising either approach's effectiveness. Resource management improvements through AI optimization involve analyzing project needs and resource availability dynamically, ensuring optimal allocation throughout project lifecycles. Risk mitigation capabilities are enhanced as AI systems can identify and address potential risks early in hybrid project workflows, often before human project managers would detect them. This augmentation creates more resilient, adaptable project management approaches that can respond effectively to changing requirements while maintaining necessary structure and control.

Real-world case studies demonstrate AI's tangible benefits in agile and hybrid project models across various industries and project types. AI tools consistently streamline workflows and automate routine tasks, leading to significant productivity improvements that allow teams to focus on higher-value activities and strategic initiatives. Improved stakeholder communication represents another key success factor, as AI facilitates clearer, faster communication among project stakeholders, ensuring better alignment and more effective collaboration throughout project lifecycles. The integration of AI in both agile and hybrid project models demonstrates remarkable real-world success and adaptability across different organizational contexts, project sizes, and industry requirements. These case studies provide compelling evidence that AI adoption delivers measurable improvements in project outcomes, team satisfaction, and stakeholder value.

Automated sustainability data collection and analysis represents a significant advancement in environmental and social impact management for projects. AI-driven data collection systems automate the gathering of environmental and social impact information, dramatically improving efficiency while reducing manual errors that often compromise data quality and reliability. Advanced AI algorithms analyze this collected data to provide accurate, comprehensive sustainability assessments for ongoing projects, enabling project managers to make informed decisions about environmental impact and social responsibility. This automation ensures consistent, reliable data collection and analysis that supports evidence-based decision-making for sustainability initiatives. The result is more effective environmental stewardship and social responsibility integration throughout project lifecycles, supporting organizational sustainability goals while maintaining project efficiency and effectiveness.

AI significantly enhances transparency and accountability in project management through two key mechanisms that build stakeholder confidence and support better decision-making. Real-time dashboards powered by AI provide clear, transparent displays of project metrics and performance indicators to all stakeholders, ensuring everyone has access to current, accurate information about project status and progress. Detailed audit trails generated by AI systems improve accountability by automatically tracking all project actions, changes, and decisions, creating comprehensive documentation that supports governance requirements and facilitates post-project analysis. These transparency and accountability enhancements build trust among project stakeholders, support better governance practices, and provide valuable insights for continuous improvement. The combination of real-time visibility and comprehensive documentation creates a robust framework for responsible project management.

Regulatory compliance and reporting requirements are significantly streamlined through AI-powered automation that reduces administrative burden while improving accuracy and timeliness. Automated compliance checks ensure that projects consistently meet legal standards and regulatory requirements by continuously monitoring project activities against applicable regulations and alerting project teams to potential issues before they become problems. Efficient documentation generation capabilities allow AI systems to produce required regulatory documentation quickly and accurately, streamlining complex reporting processes that traditionally consume significant project resources. This automation reduces compliance-related risks while freeing project teams to focus on value-adding activities rather than administrative tasks. The result is more reliable regulatory compliance, reduced administrative overhead, and improved project efficiency without compromising governance or legal requirements.

Autonomous project management systems represent the next frontier in project management evolution, characterized by three key capabilities that fundamentally transform how projects are managed. Independent planning allows autonomous systems to create comprehensive project plans without direct human input, leveraging vast data analysis capabilities and machine learning algorithms to enhance both efficiency and accuracy beyond human capabilities. Self-execution capabilities enable these systems to autonomously execute project tasks and manage workflows, significantly reducing the need for continuous manual oversight and intervention. Dynamic adjustment features allow autonomous management systems to modify project parameters, timelines, and resource allocations in real-time based on current progress data and environmental changes. These capabilities promise to revolutionize project management by handling routine tasks automatically while providing strategic insights for human decision-makers.

Chapter 13: Future Trends and Evolving Standards

The path to fully autonomous project management faces both significant technological enablers and substantial challenges that must be addressed for successful implementation. Key technological enablers include artificial intelligence, machine learning, and Internet of Things (IoT) technologies that drive innovation and enable sophisticated automated solutions capable of managing complex project environments. However, data quality challenges remain significant, as ensuring accurate, reliable data input is vital for system effectiveness but continues to present implementation difficulties. Ethical considerations play increasingly crucial roles, addressing privacy concerns, algorithmic bias, and responsible AI deployment that maintains human values and organizational integrity. System integration complexities require overcoming substantial technical and operational challenges to ensure seamless interaction between autonomous systems and existing organizational infrastructure, processes, and human team members.

The evolution toward autonomous project management systems will fundamentally transform the project manager role in three key areas that emphasize strategic value over operational tasks. The shift to an oversight role means autonomous systems will handle routine operational tasks, allowing project managers to focus on high-level oversight responsibilities and strategic decision-making that requires human judgment and experience. Strategic focus becomes increasingly important as project managers engage more deeply in strategic planning activities that guide overall project direction and ensure alignment with organizational objectives and stakeholder expectations. Enhanced stakeholder engagement becomes crucial as human project managers concentrate on communication, relationship management, and collaboration facilitation with project stakeholders. This evolution elevates the project manager role from tactical execution to strategic leadership, requiring new skills in AI system management, strategic thinking, and stakeholder relationship management.

In conclusion, artificial intelligence is fundamentally transforming the future of project management across four critical dimensions that promise enhanced effectiveness and strategic value. AI is advancing project management standards, creating more sophisticated frameworks for greater efficiency and adaptability in diverse project environments. Enhanced methodologies through AI integration enable smarter planning and execution processes that leverage data-driven insights and predictive capabilities for improved outcomes. Sustainability integration demonstrates AI's capacity to optimize resource utilization and reduce environmental impacts, supporting organizational responsibility goals while maintaining project effectiveness. The development of autonomous systems represents the ultimate evolution, automating complex operational tasks while empowering project managers to focus on strategic decision-making, stakeholder engagement, and value creation that drives organizational success in an increasingly complex business environment.